Modelling and Controlling of Behaviour for Autonomous Mobile Robots

Hendrik Skubch

Modelling and Controlling of Behaviour for Autonomous Mobile Robots

 Springer Vieweg

Hendrik Skubch
Kassel, Germany

University of Kassel, 2012
Date of Disputation 17.08.2012

ISBN 978-3-658-00810-9   ISBN 978-3-658-00811-6 (eBook)
DOI 10.1007/978-3-658-00811-6

The Deutsche Nationalbibliothek lists this publication in the Deutsche Nationalbibliografie; detailed bibliographic data are available in the Internet at http://dnb.d-nb.de.

Library of Congress Control Number: 2012953467

Springer Vieweg
© Springer Fachmedien Wiesbaden 2013

Printed on acid-free paper

Springer Vieweg is a brand of Springer DE.
Springer DE is part of Springer Science+Business Media.
www.springer-vieweg.de

Acknowledgements

I would like to thank my doctoral advisor Kurt Geihs for the amazing atmosphere he created at the Distributed Systems Group at the University of Kassel. The freedom to pursue our ideas and the friendly spirit are what makes this place so special. I would also like to thank my second supervisor, Alexander Kleiner, for the time he invested and his very helpful suggestions.

The people who influenced this work the most are Roland Reichle and Philipp Baer, who successfully completed the incredible challenge of founding a RoboCup Middle-Size team before I started work in Kassel. Without them, this work would not have been possible. At that time, I also had the pleasure of working with Thomas Weise, who provided many insights and always gladly shared his unique perspective, Michael Wagner, who is maybe the best colleague one can have, Steffen Bleul, who always managed to get our feet back on the ground, and Michael Zapf, who was always open for fruitful discussions. Together with Diana Comes and Christoph Evers, these people contributed so much to the friendly atmosphere in our group. I am also very grateful for our administrative staff, namely Thomas Kleppe, Iris Roßbach, and Heidemarie Bleckwenn, who always helped out when needed.

Carpe Noctem is now organised by Dominik Kirchner, Daniel Saur, and Andreas Witsch, who already have fascinating ideas to be implemented on the robots. The time with you was just great! Many thanks go out to the Carpe Noctem team, both current and former members, who all put in a lot of effort so that the team could become what it is now: Till Amma, Kai Baumgart, Jewgeni Beifuß, Fridolin Gawora, Tareq Haque, Janosch Henze, Timo Heumüller, Kai Liebscher, Claas Lühring, Stefan Jakob, Stefan Niemczyk, Stephan Opfer, Stefan Triller, Andreas Scharf, Jens Schreiber, Martin Segatz, Florian Seute, Daniel Walden, and Martin Wetzel. I will never forget the great time we had working, experimenting on the robots, or just hanging out.

Last but not least, I thank my parents who managed to get me this far and I am very grateful for the support and patience of Caroline Wagenaar who always encouraged me.

Hendrik Skubch

Abstract

As research progresses and capabilities of robots increase, robotic systems become viable solutions in more and more scenarios. This especially applies to multi-robotic systems, which combine the skills and functions of individual robots. Robustness, efficiency, and adaptivity are key characteristics for such future teams of autonomous mobile robots being employed in increasingly complex and dynamic scenarios. Hence, the question arises how different modelling and reasoning paradigms can be combined to describe the intended behaviour and achieve these characteristics. We present a comprehensive solution to modelling and execution of behaviour for teams of autonomous mobile robots. The proposed framework, ALICA (A Language for Interactive Cooperative Agents), combines modelling techniques drawn from different paradigms in an integrative fashion. Hierarchies of finite state machines are used to structure the behaviour of the team, such that temporal and causal relationships can be expressed. Utility functions are used to weigh different options against each other and to assign agents to different tasks. Finally, constraint satisfaction and optimisation problems are integrated, allowing for complex cooperative behaviour to be specified in a concise, theoretically well-founded manner. The system is geared towards highly dynamic environments, in which robots must act quickly, communication is unreliable, and individual robots can break down at any time. In such environments, it is imperative that agents act immediately whenever they are confronted with changing situations instead of establishing agreement beforehand. ALICA agents therefore make decisions locally and act accordingly before any communication takes place. Conflicts arising from incoherent decisions and beliefs can be reliably detected and resolved. Since ALICA works completely decentralised, no single point of failure exists.

From a modelling perspective, ALICA presents itself as a modern language in which the behaviour of a team can be modelled from a global perspective. Abstractions through hierarchical structures and program components make the inherent complexity of the topic transparent and foster reusability. The combination of state machines, utility functions, and non-linear continuous constraint satisfaction and optimisation problems is a completely new approach to describing the behaviour of a team of robots, extending the state-of-the-art.

The resulting execution layer is equipped with a novel anytime algorithm to solve the integrated constraint problems, which allows tracking of solutions over time in a dynamic environment, coordinates solutions within the team, and exploits the distributed computational power available within the team for solving.

We evaluate our approach in the robotic soccer domain, which focuses on reactivity and robustness with respect to unreliable communication. In the domain of extraterrestrial exploration, we sketch some advanced techniques for describing dynamic formations of robots. Finally, we examine the scalability of ALICA and compare the employed techniques with other state-of-the-art methods using a popular rescue scenario.

Contents

IV Assessment   193

List of Figures

List of Tables

Listings

Part I

Preliminaries

1 Introduction

1.1 Motivation

Individual robots have been a very active field of research since the robot Shakey autonomously drove around at Stanford using STRIPS (Stanford Research Institute Problem Solver) [45] to control its actions [139]. Although since then tremendous scientific advances have been made in the areas of robotics, artificial intelligence, and machine learning, only recently have robots begun to appear in every day situations. Today, robots are used as socially assistive technology in hospitals [9], employed as surgical tools [165], and are envisioned to be used in household settings in the near future [162].

Yet when robots are further integrated into modern society and economy, the demand for them to exhibit intelligent, adaptive behaviour in the dynamics of modern day to day life will grow. Apart from day to day life, one of the main goals of robotics is to provide cleverly designed machines that can be deployed in areas inaccessible to or too dangerous for humans. Such areas include rescue scenarios after catastrophic events, space exploration endeavours, or future asteroid mining [68]. Moreover, these machines can take on tasks considered too repetitive.

In these scenarios, teams of potentially heterogeneous robots are highly beneficial. A team is more robust than a single robot and can cover more ground in the same time. Further, integration of new robots is in general easier than integrating new specialised equipment into an already working robot. However, operating a team in an efficient and robust way faces new challenges compared to operating a single robot. The team has to coordinate itself, react coherently in a dynamic environment, and compensate for incapacitated members. All these issues need to be handled at runtime in an adaptive and distributed manner, since the usage of any central control mechanism voids one of the team's strongest advantages: its robustness against failing components.

Under these conditions, the creation of a recipe, or plan, to achieve a desired goal becomes a difficult task, regardless whether this recipe is generated by a planning algorithm, modelled by an expert of the respective domain, or created in an interactive planning process. It is therefore crucial that the underlying execution of said recipes exhibits the desired behaviour even in difficult situations, such as high sensory noise, unreliable communication between the team members, and dynamically changing situations.

At this juncture between the higher level recipes and the exhibited behaviour of the team, the language in which these recipes are formulated becomes the pivotal element. The language, or at least a representation of this language, must be understood by all involved parties, be it system developer, planning algorithm, or execution layer. The language should therefore feature formal semantics, such that individual tools can be verified to adhere to a common understanding of it. Furthermore, the language must be sufficiently expressive to capture the various scenarios robotic teams can be employed in.

In scenarios such as search and rescue, a robotic team has to deal with dynamically changing situations, since the environment can change abruptly, individual robots can break down, and finally communication cannot be regarded as reliable. Especially in these situations, the role of the execution layer is crucial. This component estimates for each new situation whether a selected course of action can be continued, should be repaired, or must be aborted altogether. It then selects the appropriate repair mechanism and triggers higher level components, such as a planning algorithm if need be. Depending on the situation, this might require swift reactions without communication, in other cases, communication with team members may be required to resolve conflicts. Moreover, it is this component that links the symbolic description of the intended behaviour with the lower level actuators. In other words, it grounds action symbols in physical acts. In this way, the execution layer supports cognitive abilities such as planning and learning on higher levels.

Finally, the language should be able to abstract the complexity of specific parts, such that system designers only need to concern themselves with a concrete problem at a time. That entails that the language should support separation of concerns and reusability of components.

1.2 Problem Statement

The objective of this work is to provide a comprehensive solution to modelling and controlling the behaviour of teams of cooperative autonomous robots. This solution consists of two parts: a language in which the behaviour of a robotic team can be expressed and a corresponding execution layer that understands elements of this language as programs and executes them efficiently and robustly.

In order to facilitate further tool support and allow for future embeddings into or from other languages, the modelling language must provide clear formal semantics. Depending on domain and scenario, language elements, i.e., programs, will be written by human developers or generated by planning or learning algorithms.

In contrast to many available languages and frameworks which will be discussed in Chapter 3, the framework should support modelling from a global perspective, such that the designer can focus on coordination at a high abstraction level and does not need to deal with multiple interacting individual programs. At the same time, the framework should allow for fine-grained modelling of the internal structure of the intended tasks.

As stated earlier, the proposed language must facilitate the formulation of complex behaviours in a concise and well-defined manner. Since such a behaviour always relates to domain-specific entities such as positions, the language must be able to express properties about these entities, while at the same time retain its domain-independent nature. We integrate an expressive class of constraint satisfaction and optimisation problems into the language and equip the execution layer with a corresponding solver.

This work focuses on scenarios where the inherent dynamics require fast reaction and reasoning. More specifically, the execution platform must deal with a dynamic environment where

- The situations changes continuously in real-time, i.e., while reasoning.

- Sensors are noisy, requiring a certain robustness from the execution layer.

- The environment is only partially observable, so robots cannot effectively anticipate their course of action over a longer period of time.

- Communication is unreliable. While we assume that communication is possible with a probability larger than 0, packet loss and packet delay are common occurrences.

- Individual robots can break down at any time or lose some of their capabilities. Whenever possible, the team's performance should degrade gracefully. This excludes the use of any central component.

There are certain interrelations between these properties, which create trade-offs in which this work needs to be positioned. Firstly, when dealing with sensory noise in a dynamic environment, reactivity and robustness against noise are two conflicting goals. Similarly, communication cannot be regarded as synchronous in a highly dynamic domain; therefore, received messages always refer to past situations. In particular, a sudden increase in package delay may appear as a sudden change in the environment if the clocks of the agents are not synchronised tightly enough. In general, we emphasize reactivity over robustness against noise; however, our framework can be combined with any kind of sensor processing and fusion approach which deals with noise. Thirdly, an agent breaking down is not distinguishable from packet loss, unless the agent's actions can be observed through sensory data. Finally, reactive behaviour and coordinated behaviour form two conflicting goals as well. Highly reactive behaviour cannot be achieved if communication is required for decision making. On the other hand, communication seems to be the only currently feasible way to achieve coordination. Therefore, an adaptive approach is needed, that can make swift decisions without prior communication and switches to a less reactive decision making protocol with higher coherence when necessary.

Since there is a vast number of research areas related to multi-robot systems, we exclude certain topics from this work, namely:

Sensor Fusion – Sensor fusion in heterogeneous teams of robots is discussed in detail by Reichle [131]. In all scenarios, we assume the presence of appropriate sensor fusion algorithms, such that decision making can be done on top of it.

Collaborative Systems – In collaborative systems, individuals follow potentially different goals, which can temporarily align and warrant the formation of coalitions to achieve common goals. We assume that all robots pursue the same global goal and thus form a team throughout their life-cycle.

Planning – Planning algorithms produce totally or partially ordered sequences of actions in order to achieve a specific goal. We do not devise or integrate a planning algorithm. However, we provide a language in which the output of a planning algorithm can be easily described.

1.3 Scenarios

In the following, we briefly describe the scenarios used to develop and evaluate the result of this work. Note that our approach is not limited to these domains and can easily be transferred to others. This is emphasized by the clear distinction between domain-specific entities and general entities made in our approach. Hence, the domain description is easily exchangeable.

1.3.1 RoboCup

RoboCup is a multi-national research endeavour fostering advances in robotics, artificial intelligence, and related fields. At its core is an annual competition in different leagues and a vision:

> "By mid-21st century, a team of fully autonomous humanoid robot soccer players shall win the soccer game, comply with the official rule of the FIFA, against the winner of the most recent World Cup." – The RoboCup Website [137]

With this vision, RoboCup also commits to a main research scenario, namely robotic soccer. Since its first competition in 1996, RoboCup has been expanded and now targets other domains as well, such as rescue robotics and robots in household environments. However, the simple, yet surprisingly challenging soccer domain is still its primary focus. The advantages of soccer over other domains as a research domain are plentiful: Soccer is a well-known sport and therefore requires little explanation and holds interest for the general public. It also provides a test-bed where completely different approaches can be evaluated against each other in a natural way: by competition. Moreover, in soccer, the skills of each individual player are as important as teamwork among the players. Thus, both must be pursued in order to form a successful robotic soccer team. Finally, soccer is a fast-paced game, requiring a high degree of reactivity and speed. This requirement differentiates RoboCup from typical laboratory settings and must be accounted for in all aspects of a participating robotic team.

When referring to the RoboCup domain in this work, we focus on a specific RoboCup league, namely the middle size league (MSL). In the MSL, teams of five robots compete against each other on a field of $18\,m \times 12\,m$. A game consists of two half-times, each lasting $15\,min$. The robots use a normal FIFA soccer ball to play. During a game, each robot acts completely autonomous; human interaction is strictly forbidden. Currently, competing robots are up to $80\,cm$ high and weigh no more than $40\,kg$. Most robots use a holonomic drive to move about and a separate kicking device to kick the ball over longer distances. The robots of a team communicate via wireless LAN. The game is very dynamic; robots can reach speeds of $5\,m/s$ and accelerate the ball to about $12\,m/s$. Given that the typical sensory range of a robot is no more than $10\,m$ under ideal conditions and often less than $6\,m$, reaction speed plays a pivotal role in the performance of a robot. But reacting is not enough within a team, the robots also have to act and react *coherently*. In this work, we explore possibilities to maintain swift reaction time while coordinating the team to establish coherent actions.

1.3.2 Exploration

One scenario in which the usage of autonomous mobile robots is highly beneficial is exploration. The environment encountered in exploration missions is typically either difficult to reach in the first place or hazardous to humans. In the case of extraterrestrial exploration, both are the case. Moreover, the large distances between celestial bodies render teleoperation difficult or outright impossible. Therefore, it is crucial to endow exploring robots with an appropriate level of autonomy.

Such scenarios are tackled by the research project IMPERA, which is undertaken by DFKI[1] Robotics Innovation Center together with the Distributed Systems Group of the University of Kassel and coordinated by the DLR[2]. Following the idea that a team of robots is more robust and fault tolerant than a single robot and that a team can even accomplish tasks such as exploration faster than a single robot, IMPERA investigates methods to coordinate teams of robots in extraterrestrial environments.

While the environment of Mars or the Moon, for instance, are far less dynamic than a soccer game, the other domain features we discussed in Section 1.2 are to be considered. However, the robotic team itself can induce certain dynamics in the environment by interaction.

The approach we present in this work serves as a basis for IMPERA to control and coordinate robots during extraterrestrial missions. Most notably, IMPERA will extend this work with state-of-the-art planning algorithms.

1.3.3 Rescue

The potential tasks, which robots can fulfil in rescue scenarios are manifold. Robots might be employed to search large areas for survivors, provide communication networks, clear rubble, and enter unstable buildings. In the context of this work, we investigate a simple scenario, based on the RoboCup Rescue Simulation League. In this scenario, agents are tasked with extinguishing fires that break out in multiple places in a city. We simulate this benchmark problem using RMAS-BENCH by Kleiner et al. [86]. This domain requires to coordinate many more agents than exploration or soccer scenarios typically feature.

---

[1]   Deutsches Forschungszentrum für künstliche Intelligenz – German Research Center for Artificial Intelligence
[2]   Deutsches Zentrum für Luft- und Raumfahrt – German Aerospace Center

1.4 Approach

As previously mentioned, the proposed solution consists of a language and a corresponding execution layer. In order to provide an abstraction between the described behaviour and the concrete robots at hand during execution, a two tiered mapping between robots and tasks within the described behaviour is used. Agents, or robots, are mapped onto roles based on their individual capabilities regarding specific actuators or sensors. Roles are then used to determine the tasks an agent can take on and how well it can perform them. This mapping closely follows the ideas described by Wooldridge et al. [182]. Thereby the team behaviour can be specified independently of the concrete robots available during runtime and the team can be specified independently of the problem it should solve later on.

The program encoding the team behaviour is a hierarchical structure consisting of multiple finite state machines. Each state in such a machine can contain sub-programs, which are meant to be executed by all agents inhabiting that state. At each level, conditions and utility functions are used to determine which state machines are executed by which agent. Evaluation of these conditions and functions is done locally by each agent, allowing for highly reactive behaviour. The hierarchical structure hides the complexity of lower levels at each level.

This base language is extended by a conflict detection and resolution mechanism, which temporally switches the decision protocol employed by electing a leader in order to resolve the detected conflict. The leader then commands the rest of the team only with respect to the decision in conflict. After the conflict is resolved, the team switches back to its original, more dynamic decision making mode.

Finally, constraints are used to describe target values that robots should pursue. This results in dynamic constraint optimisation problems, which are solved by the team members during runtime. We provide an efficient anytime solver for non-linear continuous constraint optimisation problems for this task. This solver is able to track solutions as the environment and, correspondingly, the optimisation problem changes. Moreover, the team is cooperating in solving harder problem instances, such that it can find a solution faster than a single agent can. Finally, the solving algorithm enables the team to coordinate its solutions, i.e., come to a coherent result. The execution layer manages the set of active constraints, allowing any component to query variables for their values.

The benefits of incorporating constraints as declarative descriptions into the language are plentiful.

- It greatly simplifies the modelling task and reduces the number of necessary atomic behaviours.

- Constraints are easy to extend and combine as their mathematical relations are well understood. For instance, constructing the intersection of two constraint satisfaction problems is straightforward.

- In comparison to imperative programs, far fewer special cases need to be addressed.

- The direct correspondence of the implementation to the mathematical description enables the generation of constraints by planning and learning algorithms.

- Most importantly, constraints offer a way to ground symbolic behaviour descriptions in numeric values which can be passed to lower level components such as motor controllers.

Naturally, there is also a drawback. A general purpose solver is always outperformed by problem-specific solutions in terms of efficiency. Hence, an imperative program designed to tackle a specific task can always be more efficient than a constraint-based solution to the same problem.

1.5 Contributions

The main contribution of this work is a comprehensive solution to describing the behaviour of robotic teams. The combination of paradigms presented here constitutes a novel way to model the cooperative behaviour of multi-robot systems. The approach, coined ALICA (A Language for Interactive Cooperative Agents), encompasses description of capabilities, role allocation, task description similar to hierarchical state machines, task allocation based on utility functions, explicit coordination through synchronised transitions, implicit coordination in a broadcast-and-compute fashion, and conflict detection and resolution by switching the decision making protocol on the fly. Finally, non-linear constraint satisfaction problems are integrated in order to allow reasoning over domain-specific entities such as positions or configurations of joints. We see this comprehensive combination as the largest contribution of this work.

The implementation has been successfully employed by the RoboCup MSL team Carpe Noctem since 2009 and is currently being used in the research project IMPERA. The source code is available under a BSD-based open source license.[1]

On the theoretical level, we provide new insights into hierarchical task allocation leading to two different allocation schemes with different requirements on the task

---

[1]   http://ros.org/wiki/cn-alica-ros-pkg

structure. Furthermore, an anytime constraint optimisation solver is presented, which extends state-of-the-art approaches for non-linear continuous satisfaction problems. Most notable are its abilities to track solutions over time, coordinate solutions within the team, and distributively solve problems while retaining reactivity of the individual robot.

1.6 Structure of this Work

This document is divided into four parts. The remainder of the first part consists of an introduction to the foundations of this work in Chapter 2, followed by a discussion of related work with respect to our approach in Chapter 3.

Part II presents the propositional core language of ALICA in detail. After discussing its syntax in Chapter 4, the semantics is introduced step-by-step in Chapter 5. Task allocation and rule-based execution are emphasized in this part of the thesis, as they form the foundations for the following extensions.

Chapter 6 extends the basic semantics with conflict detection and resolution. A domain-independent conflict detection scheme is derived, which allows reliable detection of persistent conflicts within team coordination. Afterwards, we present a conflict resolution approach based on local leader election. Part II ends with a discussion of the underlying software architecture and implementation details in Chapter 7.

Part III presents an extension of the propositional language by incorporating variables and constraints over them. Correspondingly, the syntax and operational semantics are extended in Chapter 8 to accommodate for the additional language elements. Further, in Chapter 9 a suitable constraint solving algorithm is derived and discussed in detail. The rest of the chapter explains how solutions can be tracked over time and how coordination in this non-propositional case can be achieved.

Part IV is concerned with the evaluation and discussion of the presented solution. In Chapter 10, we present thorough evaluation results. Using scenarios drawn from the domains robotic soccer, exploration, and rescue, we highlight robustness under poor network conditions, robustness under noise, modelling capabilities, scalability, and applicability for different scenarios. We conclude in Chapter 11 and discuss possible future work.

1.7 Conventions

For the sake of brevity, we assume the following notational conventions in this
work:

- Subtraction of finite sets is denoted by $-$:

$$A - B \stackrel{def}{=} \{a \mid a \in A \wedge a \notin B\}$$

- Free variables in formulae are universally quantified unless otherwise stated.

- The following abbreviations are used in first-order formulae:

$$(\forall x \in S)\phi \stackrel{def}{=} (\forall x)x \in S \rightarrow \phi$$

$$(\exists x \in S)\phi \stackrel{def}{=} (\exists x)x \in S \wedge \phi$$

- By $\mathrm{img}(f)$ we denote the image of function f in the usual sense.

- 2^S denotes the power set of set S.

- In cases where we explicitly refer to free variables in terms or formulae, we
  use $\mathrm{vars}(p)$ to denote the set of free variables of term or formula p.

2 Foundations

In this chapter, we briefly explain the main theoretical foundations of this work. Section 2.1 introduces the concept of agents, followed by a classification of different multi-agent systems in Section 2.2. Section 2.3 illustrates the main teamwork theories in the literature. Finally, Section 2.4 gives a short overview on constraint programming.

2.1 Agents

The concept of agents has been used in various settings. Agents were explored by researchers as a software technology entity (e.g., [115]), as means to simulate the behaviour of ecosystems (e.g., [61]), and as a concept of artificial intelligence, differentiating between an environment and an interior.

> "An agent is anything that can be viewed as perceiving its environment through sensors and acting upon that environment through actuators."
> – Russell and Norvig [139, p. 32]

This is probably the most generic agent definition in the literature. An agent typically implements a control loop as depicted in Figure 2.1, although the names and numbers of internal steps vary between models.

In the first step, sensory input is processed, the result is used in a reasoning step, which yields a decision for actions. Executing this action will in turn modify the environment, which then leads to different input. There are various ways to classify agents, depending on how these steps are implemented, what kind of internal model is used and with what kind of environment the agent can deal with. An excellent overview is given by Russell and Norvig [139]. Here, we limit the discussion to three different agent models, namely rational agents, BDI-agents, and reasoning agents.

Rational agents try to maximise their expected performance measure. That is, they behave in a decision-theoretically optimal sense, provided appropriate information about possible rewards and probabilities is available.

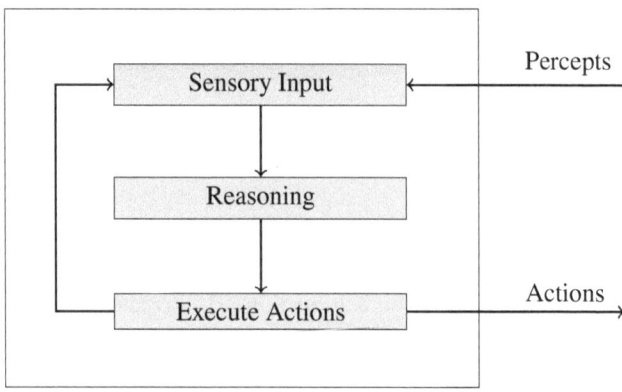

Figure 2.1: Agent Execution Loop

BDI-agents follow the BDI model by Bratman [12]. BDI concentrates on a practical way to perform reasoning, imbued with particular mental attitudes, namely: *Beliefs*, *Desires*, and *Intentions*. Beliefs represent the informational state of the agent about the world (including itself and other agents). The desires of an agent represent objectives or situations that the agent would like to accomplish or bring about. Intentions represent the deliberative state of the agent: what the agent has chosen to do. This practical approach was highly successful and led to a multitude of languages based on the BDI model.

Reasoning agents follow a very classical view of AI, where reasoning is mapped onto deductions in some formal logic. Research in this area focuses on how knowledge can be represented to allow for powerful reasoning techniques to be used. Foundational work in this area dates back to McCarthy and Hayes [100]. Two of the most influential calculi for reasoning agents are the situation calculus [134] and the fluent calculus [72, 166].

Our work is compatible with all of these agent models, although ALICA is strongly inspired by the BDI model, and can be seen as a BDI language. In the following, we use the words agent and robot interchangeably.

2.2 Multi-Agent Systems

Based on the notion of agents, systems comprising multiple agents were, and still are, subject to intensive research. Such systems can be classified in various ways, e.g., based on the abilities of the individuals, the organisational metaphors used, or whether the agents can be considered homogeneous or heterogeneous. An excellent overview over the field is given by Wooldridge [181]. One of the most important aspects for this thesis is the level of cooperation. A system can be classified as:

Cooperative Meaning that all agents try to achieve the same global goal. They will therefore cooperate in any way that is deemed beneficial to this goal.

Collaborative Agents do not necessarily try to achieve the same goal, but goals that can be compatible. They will therefore form coalitions whenever there individual goals align. Compared to cooperative systems, the organisational structure is much more dynamic, since teams can form and disband continuously within the system.

Neutral Agents have different goals, and ignore each other. Any form of cooperation happens purely by chance.

Antagonistic Agents have goals that are in direct competition with each other, that is one agent obtaining its goal entails that no other agent is able to do so.

Naturally, these categories are not strict and actual systems can feature elements of each category. In robotic soccer for instance, each team forms a purely cooperative system and each team player is completely altruistic. However, the multi-agent system that is formed by the two opposing teams during a match is, of course, antagonistic.

In our work, we only consider purely cooperative systems. While our main scenario contains an opponent team, we are not interested in analysing the overall behaviour of soccer playing robots, but instead want to model the behaviour of a single team. In other words, we treat the opponent team as part of the environment, since we cannot control their actions.

Multi-agent systems which feature a large number of typically homogeneous robots with fairly simple individual cognitive and communicative abilities are investigated by the fields of swarm intelligence and swarm robotics. Swarm-based approaches are often inspired by biological systems, e.g., [119]. Modelling a set of robots as a team on the other hand assumes that the individual is already capable of relatively complex reasoning. In Section 2.3 we will discuss the main theories relevant to cooperative teams.

Besides swarms and teams, there are plenty of different possibilities to organise multi-agent systems. Horling and Lesser [73] give a thorough overview of the different paradigms investigated in the literature:

Hierarchy In a hierarchy, agents at higher levels have a more global view than agents at lower levels. Typically, agents only communicate with those directly connected to them through the tree-like structure. Data is communicated upwards, while commands are passed downwards. While the hierarchical structure is appealing in its simplicity and its relationship with hierarchical goal or task trees, it is rather rigid, cannot adapt well, and can feature single points of failure.

Holarchy Holarchies are nature-inspired structures, similar to hierarchies, where a group is formed at each level out of a certain kind of parts and these parts again are groups of parts at a lower level of abstraction. At the lowest level, a group consists of agents. Thus, holarchies can be seen as less strict hierarchies, allowing more communication between individuals and more autonomy.

Coalition Coalitions follow the idea of collaborative agents noted above. Here, agents form relatively short-lived, flat groups in order to achieve aligned goals.

Team Teams consist of cooperative agents that work together to achieve a common goal. In contrast to coalitions, they form a purely cooperative system.

Congregation Congregations are long-lived flat hierarchies of agents. In contrast to teams, they do not have a single goal, but form in order to combine complementing capabilities.

Society Agent societies are long-lived open systems, such as an electronic market. Agents pursue different goals, have heterogeneous capabilities, and interact with each other through various channels. The society imposes a set of constraints individuals must adhere to, commonly named social laws or norms.

Federation In agent federations, potentially complex groups of agents are each represented by a single distinguished member of the group, which is in charge of communicating and interacting with the representatives of other groups.

Market In contrast to societies, the whole interaction process in markets is modelled after commerce, i.e., they buying and selling of goods, placing bids, and offering services. The situation is typically competitive; that is, the individual goals conflict with each other.

Matrix Matrix-based agent organisations allow the definition of multiple dimensions of authority. The individual agent has to be equipped with a sufficient amount of autonomy in order to deal with the potential local conflicts than can arise from dealing with different authorities.

Compound The last paradigm basically combines different organisational structures for different purposes, such as data-flow, control, discovery, etc.

Since we are most interested in small scale groups of robots following a common goal, the approach presented in this work follows the characteristics of a team-based multi-agent system.

2.3 Teamwork

Teamwork among intelligent agents has been extensively analysed in the literature. Today, most approaches to modelling teamwork in agent languages are based on at least one of the two following prominent theories:

Joint Intentions Theory The Joint Intentions Framework [28, 97] is a theoretical framework founded on BDI logics. The framework focuses on a team's joint mental state, called a **joint intention**. A team jointly intends a team action if all team members are jointly committed to perform an action while in a specified mental state.

In order to enter a joint commitment, the team members have to establish appropriate mutual beliefs and individual commitments. Although the Joint Intentions Theory does not mandate communication and several techniques are available to establish mutual beliefs about actions from observations (see for example [75]), currently communication seems to be the only feasible way to attain joint commitments. A very interesting key aspect of the Joint Intention Theory is the commitment to attain mutual belief about the termination of a team action. This helps to ensure that the team stays updated about the status of the team actions. This behaviour is achieved by enforcing that agents committing to a joint intention also commit to inform their team about any relevant failures or premature terminations. Joint intentions and joint commitments provide a basic framework to reason about coordination required for teamwork as well as guidance for monitoring and maintaining team activities. However, a single joint intention for a high-level team goal is not sufficient to model team behaviour in detail and to ensure coherent teamwork.

Shared Plans Theory   In contrast to Joint Intentions, the Shared Plans Theory [63, 64] employs hierarchical structures over intentions, thus overcoming the shortcoming of a single Joint Intention for complex team tasks. The Shared Plans Theory is not based on a joint mental attitude but on an intentional attitude called **intending that**, which is very similar to an agent's normal intention to perform an action. However, an individual agent's 'intention that' is directed towards its collaborator's action or towards a group's joint action. 'Intention that' is defined via a set of axioms that guide an individual to take actions (including the communication), that enable or facilitate its team-mates, sub-team, or team to perform assigned tasks.

A SharedPlan for a group action specifies beliefs about how to do an action and sub-actions [63, 64]. The formal model captures intentions and commitments toward the performance of individual and group actions. A collaborative plan is composed of a **mutual belief**, of a (partial) recipe, individual **intentions to** perform the actions, individual **intentions that** collaborators succeed in their sub-actions and individual or collaborative plans for sub-actions. With the concept of actions and sub-actions the Shared Plans Theory describes a hierarchy of plans to reach a common goal. This is also the main difference between the Joint Intentions Theory and the Shared Plans Theory; the Shared Plans Theory describes the way to achieve a common goal whereas the Joint Intentions Theory describes only this common goal. However, the lack of principles like joint intentions and joint commitments results in limited possibilities to reason about team coordination and team activities.

The two theories Joint Intentions and SharedPlans have been extensively used to examine and describe teamwork. In our approach, we will draw from both, although the relationship to SharedPlans is most apparent due to the structural similarities between collaborative plans and ALICA plans in execution.

2.4 Constraint Programming

The paradigm of constraint programming advocates a declarative description of problems, which are then solved by appropriately chosen constraint solvers. A recent overview is given in Rossi et al. [138]. A very general framework has been proposed by Frühwirth [49, 50], which defines *Constraint Handling Rules* (CHRs) as a unified way to express constrains. Slightly simplified, a constraint handling rule is of the form Head <==> Guard | Body, meaning, that if some of the constraints currently imposed unify with Head, and the Guard evaluates to true given the unification without modifying variables in the Head, the Head is replaced by Body. This system has been integrated into the constraint programming framework ECL^iPS^e [2]. CHRs can be used to propagate and simplify constraints, however, they need to be combined with a search method in order to identify solutions to the constraint satisfaction problem encoded. Such a search typically makes assumptions about the variables by posting further constraints and backtracks once an assumption leads to a detectably unsatisfiable constraint. Backtracking can be extended to backjumping, where multiple steps are retracted at once in order to search a more promising region of the search space earlier [124, 26].

Constraint satisfaction in a Boolean domain has been shown to be NP-complete [29, 143]. Later, the same could be shown for general constraint satisfaction problems over finite domains [43]. While some tractable subclasses were identified [27], these hardly fit the requirements of robot control, especially since many different problems need to be formulated.

The class of problems spanning over continuous domains is even more interesting from the perspective of robotics research, as many values that appear in robotic domains are real-valued, such as positions or angular states of joints. In general, the problem of solving constraint-based mathematical models over the real numbers is undecidable [135]. However, solutions can be approximated. Finding intervals that approximate solutions of reasonable precision is NP-hard [10]. The most general case of a constraint satisfaction problem corresponds to a first order formula for whose free variables a solution is to be found, and this problem is undecidable, due to satisfiability being undecidable in first order logic in general.

In order to capture more complex problems, Jónsson and Frank [80] introduced dynamic constraint problems, where individual constraint problems are linked in a sequence. Neighbouring problems can be obtained from each other by restriction or relaxation. They propose a reasoning technique based on procedures to solve such systems. Thereby, the resulting system can solve problems efficiently, provided corresponding procedures are given. In this way, they relax the requirement that all variables need to be known before solving a system, and can consider problems where the number of variables are unknown in advance. This way, unbounded planning tasks become representable as constraint satisfaction problems. Similarly, Nareyek [105] represented planning as a constraint-based local search in a space of graphs, where each graphs represents a possible plan.

More recently, distributed constraint optimisation problems have been used to control the behaviour of MAS systems. This field is described in detail by Petcu [121]. In this setting, each agent owns and controls a constraint optimisation problem, which relates to the problems possessed by other agents through some shared variables. The agents obtain a global solution by interleaved local solving and message exchange. Typically, no agent has a global view on the problem.

In summary, constraint programming constitutes a very concise and mathematically well-founded way to express problems. This is the main motivation to integrate constraint programming techniques into our solution. However, the problem classes that occur in the domains we consider together with the inherent dynamics of these domains have not yet been tackled under soft real-time considerations from a constraint programming point of view.

3 Related Work

There are various approaches to describing the behaviour of agents, from the famous STRIPS planning language [45] and the first formalisms following the BDI model by Bratman [12], to modern languages such as 2APL by Dastani et al. [35]. In the following, we will give a short overview of this vast field and discuss different approaches in relation to our language.

3.1 Action Calculi

STRIPS [45] was one the first formalisms to allow agents to reason about their actions. Although STRIPS was limited to describing actions using propositional preconditions and effects only, it was surprisingly effective. This work and the foundational questions raised by McCarthy and Hayes [100], such as the frame problem, led to the development of the situational calculus [133, 134]. A family of programming languages were developed based on this theoretical foundation. Its earliest member, GOLOG by Levesque et al. [98], was followed by numerous variants such as ConGolog [37], which allows for concurrency, reaction to exogenous actions, and interrupts. Later on the prominent IndiGolog [57] added support for planning and searching.

Meanwhile, a second calculus was developed by Thielscher [166] based on the work by Hölldobler and Schneeberger [72], the fluent calculus. It focuses more on the dynamic facts that describe the environment than on the actions that change them. Thielscher [167] developed a programming framework based on the fluent calculus, called FLUX. FLUX is mainly concerned with the representation of partial knowledge and updates of this knowledge. Both are achieved through constraints.

The event calculus by Kowalski and Sergot [88] is another highly promising approach, as it allows reasoning with time intervals, while the previously mentioned calculi only consider discrete states. Details can be found in [151]. Unfortunately, to our knowledge, there is no complete agent-oriented language based on the event calculus yet.

All these languages focus on the representation of the agent's knowledge, the effect of actions and how reasoning can be done based on this knowledge. In contrast, ALICA focuses on the representation of strategies that are meant to solve certain problems or deal with specific situations. Moreover, ALICA is a team-centric language, while the languages above typically focus on single agent scenarios. Thus, from the perspective of action calculi, ALICA can be seen as a possible program representation that is combinable with, for instance, a GOLOG or FLUX dialect. Similarly, from the perspective of ALICA, action calculi-based languages fill the gap that is needed to reason about behaviour, as ALICA intentionally leaves the representation of the environment open.

3.2 BDI Languages

The BDI model by Bratman [12] (see Section 2.1) led to a set of successful agent languages. While ALICA can be seen as a BDI language, it lacks an explicit representation of desires or goals, since it focuses on plans as intentions. In the following, we discuss some of the most influential BDI languages in relation to ALICA:

3APL   is an agent oriented programming language [71, 34] aiming at modelling cognitive agents and high level control of cognitive robots. ALICA shares many concepts with 3APL, e.g., the definition of the belief base and the interpretation of goals as 'goals-to-do', which are not described declaratively but via plans that are directed towards achieving a goal. However, in contrast to ALICA, 3APL also facilitates explicit specification of goals. It introduces rule sets and beliefs to allow reasoning over both, goals and plans. Moreover, ALICA defines its operational semantics through a transition system much in the same way as 3APL. In fact, 3APL distinguishes between a transition system for the pure language elements and a transition system for the meta-language to specify the control structures of an agent. In ALICA, we do not provide this distinction, and thus, the two transition systems are merged. Although 3APL implementations support communication in a FIPA[1] [47, 46] compliant manner, explicit multi-agent plans as supported by ALICA cannot be expressed in 3APL. ALICA also extends the notion of runtime substitutions, which are part of an agent configuration in 3APL, by replacing them with a more expressive constraint store.

2APL   by Dastani et al. [35] is a successor of 3APL, featuring various programming constructs such as exception handling, repair mechanisms, and language interfaces. However, 2APL does not feature any way to model multi-agent plans

---

[1]   Foundation for Intelligent Physical Agents

from a global perspective. Instead, single agent plans need to be devised which interact with each other through explicit messages.

AgentSpeak(L)   allows BDI agents to be specified similar to logic programs [129]. Rao identified a gap between implemented BDI systems and the theory.[1] Rao tried to overcome this shortcoming by introducing the AgentSpeak(L) which abstracts an implemented BDI system. AgentSpeak(L) is a programming language based on a restricted first-order language with events and actions. Unfortunately, the modelling of multi-agent plans is not possible in AgentSpeak(L). Interestingly, AgentSpeak(L) can be simulated by and thus embedded into 3APL [69].

KARO   is not a programming language, but an agent logic based on dynamic logic. However, Hindriks and Meyer [70] proposed a programming language that directly relates to the logic. We argue that the modalities of dynamic logic are only of limited use in robotic scenarios, where actions happen concurrently and extend over time intervals. As such, robotic scenarios are potentially easier to describe in theories working with time intervals, such as the event calculus [151] mentioned above.

3.3 Plan Execution Languages

Viewed from a more practically motivated perspective, plan execution languages do not follow a rational agent approach, instead they typically provide an execution layer for plans and a language to specify them. While this is also what ALICA aims at, typically they do not provide extensive means to repair failed plans in the sense of BDI languages. Furthermore, they do not integrate coordination as tightly as ALICA and languages such as STEAM [163] (see Section 3.4).

PLEXIL   One of the most prominent plan execution languages is NASA's PLEXIL [40]. PLEXIL is completely geared towards deterministic execution, yielding itself well to its primary field of usage, namely semi-autonomous space vehicles, such as satellites. It maintains tasks as a tree-based structure and provides synchronous execution semantics. All nodes in the tree are executed in parallel, and can communicate via shared variables. PLEXIL does not provide any multi-agent semantics. Its fine grained programming metaphors allow assignments to

---

[1]   Independently, Wooldridge [181] identified the same as the ungrounded semantics problem.

variables and calls to library functions. In contrast, the smallest executable elements in ALICA are behaviours, which are essentially Turing-complete programs.

SMACH   Another noteworthy framework that falls into the category of plan execution languages is SMACH by Bohren and Cousins [11], a library for task level coordination and execution in ROS [127]. It understands itself as a mid-level task execution between a higher-level planning system and low-level action primitives. SMACH mainly supports hierarchical state machines, although other execution policies are possible.

The practical internal structures used by SMACH, namely hierarchical state machines, which allow, among others, concurrency and service calls within the ROS framework, make SMACH among all the languages mentioned here probably the language that most closely resembles the internal structure of an ALICA program. However, SMACH only considers single robots and indeed provides coordination metaphors only among components of a single system. ALICA, on the other hand, focuses on the robotic team and instead provides hierarchies of *sets* of state machines. Additionally, ALICA incorporates task allocation by means of utility functions and constraint-based modelling of team behaviour.

XABSL   A prominent behaviour modelling approach that can be seen as a plan execution language was developed by Lötzsch et al. [99], namely the language XABSL. XABSL describes agent behaviour through a hierarchical structure of states and options. Zweigle et al. [187] expanded this approach with a petri-net-like structure to capture multi-agent interaction. The graphically modelled *XPlM Nets* were compiled into XABSL trees. Although the general idea of a hierarchical interaction net is central to ALICA as well, there are fundamental differences to XPlM Nets. Most importantly, ALICA describes agents using capabilities and implements a double-layered abstraction between plans and agents through the use of roles and tasks. This allows for easier definition of complex cooperative plans, which rely on heterogeneous capabilities of the involved agents. Moreover, ALICA combines the state-based description of behaviour with utilities for task allocation and constraints. The former allow for highly dynamic, yet stable adaptations of the team's behaviour, while the latter extend the language with a fraction of first-order logic able to express complex relations over the real numbers. Hence, ALICA exceeds simple hierarchical state-based approaches.

3.4 Teamwork

The teamwork theories Joint Intentions and Shared Plans discussed in Section 2.3 led to a number of implementations. Here we briefly highlight the most prominent ones.

GRATE*   The system GRATE* by Jennings [78] is based on the Joint Intention Theory. GRATE* provides a rule-based modelling approach to cooperation using the notion of Joint Responsibilities, which in turn is based on Join Intentions. However, GRATE* is geared towards industrial settings in which both agents and the communication between them can be considered to be reliable. Thus it uses central concepts to organise the establishment of joint actions and uses extensive communication protocols before the corresponding actions are executed.

STEAM   STEAM (Shell for Teamwork) [163, 125] builds on both Joint Intention Theory and Shared Plan Theory and tries to overcome their shortcomings. Based on joint intentions, STEAM builds up hierarchical structures that parallel the Shared Plan Theory as described in the previous chapter. Hence, STEAM formalises commitments by building and maintaining joint intentions and uses Shared Plans to formulate the team's attitudes in complex tasks.

ALICA is very similar to and borrows a number of ideas from STEAM and thus also from the Joint Intentions Theory and from the Shared Plans Theory. Just like STEAM, ALICA builds hierarchical structures of team plans that cover the collaborative behaviour of whole teams and sub-teams, provides mechanisms to assign agents to (sub-)teams, and identifies the need for tracking of actions performed by teammates. ALICA also draws from the Joint Intention Theory, in particular in the definition of Synchronisations (see Section 5.13) and in capturing the need to communicate failures. However, in ALICA, failures are communicated implicitly in periodic messages (see Section 7.5).

In contrast to STEAM, ALICA agents in general do not establish joint intentions before acting towards a cooperative goal. Instead, each agent estimates the decisions of its teammates and acts upon this estimation. Conflicting individual decisions are detected and corrected using the periodically communicated internal states of teammates. Although STEAM provides approaches for selective communication and tracking of mental attitudes of team-mates, we argue that for highly dynamic domains and time-critical applications the strict requirement to establish or estimate a joint commitment before a joint activity is started has to be skipped. In ALICA, agents decide and act until contradictory information is available, which seems to be much more suitable for such applications. Nevertheless,

ALICA provides language elements to enforce an explicit agreement, resulting in a joint intention, for activities that require time critical synchronisations, such as cooperative lifting of an object. Also the assignment of agents to teams and teams to operators (which encapsulate the actual team-behaviour) done by STEAM seems to be too static for highly dynamic domains. For example in a soccer game, a robot that is assigned as defender should also be able to take over the tasks of an attacker if it obtains the ball and the game situation seems to be promising to do so. In order to facilitate such behaviour, we provide a slightly different definition of roles and incorporate the concept of tasks and preferences towards tasks. Unlike STEAM, ALICA also does not rely on the concept of a team leader, which STEAM assumes for different purposes. Instead, ALICA only reverts to leader-based decision making if a persisting conflict is detected. Thus it switches the coordination protocol during runtime.

The project "Machinetta" [142] is based on STEAM. In order to provide a lightweight and portable implementation of the teamwork framework, Machinetta uses the concept of proxies to build a reusable software package that encapsulates the teamwork model. Each proxy works closely with a single domain agent, representing that agent in the team.

While STEAM bridged the gap between cooperation theory and practice, it is not a complete implementation for a working robotic team. It provides mechanisms to reason about or to establish teamwork, but does not go into detail about the description of the internals of plans or operators. STEAM and its implementation TEAMCORE [126] just assume reactive or situated plans, do not provide support to really 'program' plans with regard to sequential and/or parallel actions, and do not really specify the internal control cycle of an agent. In this context, agent programming languages have inspired the design of ALICA, most notably 3APL [71] and its successor 2APL [35]. Tambe et al. later on defined a framework for team-oriented programming [164] based on TEAMCORE. Here, individual agents, programmed in different languages coordinate themselves via TEAMCORE proxies. This approach lends itself very well to coordinating heterogeneously programmed agents working together over distances. However, it does not address the problem of a team of robots confronted with a dynamic domain any further.

CAST   CAST [183, 184] (Collaborative Agents for Simulating Teamwork) is a teamwork framework based on the Shared Plans Theory. CAST focuses on flexibility in dynamic environments and on proactive information exchange enabled by anticipating what information team members will need. Petri Nets are used to represent both the team structure and the teamwork process, i.e., the plans to be executed. This representation avoids the computational complexity of reasoning with beliefs and is very similar to the finite state machines employed by ALICA. Furthermore, the dynamic role selection employed by CAST bears similarities to the dynamic task allocation rule we will discuss in Section 5.13.

However, the representation of roles in CAST is less expressive than the two-tiered abstraction based on agents, roles, and tasks used in ALICA. Furthermore, CAST cannot compensate for team members breaking down. Finally, in cases where the number of agents that can take on a task is bounded (XOR operator), CAST requires communication before an action can be taken.

ALLIANCE The ALLIANCE architecture by Parker [118] aims at fault tolerance in a cooperative team of mobile robots. They established a framework in which robots coordinate and adapt to faults without any central control. Similar to ALICA, ALLIANCE relies on periodic broadcast communication to achieve its goal. In contrast to ALICA, ALLIANCE does not feature the notion of high-level team-oriented plans, in which complex coordinated behaviour can be expressed. Instead, ALLIANCE defines sets of behaviours, of which one is executed while the other sets hibernate. Using a concept based on impatience and acquiescence values, robots dynamically select the appropriate set. Each group is controlled by a *motivational behaviour*, which is interrelated with the motivational behaviours of other groups. While this approach can cope with dynamically changing situations, this modelling perspective does not lend itself very well to complex plans which may feature a high number of causal and temporal dependencies. In particular, transforming a partially ordered plan generated by a planning algorithm into a corresponding ALLIANCE program seems hardly feasible.

Reis et al. [132] identified the trade-off between reactivity and cooperation when coping with the necessity to coordinate in a dynamic, noisy, real-time environment that might feature adversaries. In order to deal with this trade-off, they propose to differentiate between what they call strategic situations and active situations, essentially introducing a flat hierarchy. ALICA extends this in a sense by allowing arbitrarily deeply nested plans, so that the degree of reactivity can be tuned individually at each level using thresholds and similarity measures for task reallocation (see Section 5.8). Furthermore, Reis et al. distinguish between positions and roles. Positions dictate the physical place an agent has within a formation, while roles determine how they behave. Their approach was successfully used in the RoboCup simulation league.

However, the role exchange algorithm DPRE (Dynamic Positioning and Role Exchange) they employ only allows role switching between two agents and does not consider a completely new allocation. It is unclear what happens if two role switches collide and the corresponding utility measures interrelate as DPRE uses potential utility gains when deciding whether or not to exchange roles with another agent. The approach does not transfer well from the soccer domain since it considers all agents to be homogeneous and integrates formations and positions very tightly into the framework. Finally, their approach does not feature any plan-internal structure, nor any repair mechanisms other than role exchange.

None of the teamwork approaches discussed here feature behaviour modelling via constraint satisfaction and optimisation problems. To the best of our knowledge, ALICA is the first comprehensive framework for teamwork in robotic domains, in which such declarative description is possible as an completely integrated language element. Since ALICA solves problems formulated in this way dynamically during runtime, coordinates solutions robustly within the team, and utilises the team's combined computational power to solve hard problems if necessary, this feature is the most important contribution of ALICA to teamwork approaches.

3.5 Task and Role Allocation

There has been extensive research on the topic of task and role allocation; discussing them in any comprehensive way would exceed the boundaries of this work. Campbell and Wu recently provided a well-researched overview in [20, 21]. Gerkey [55] identified different problem classes in which task allocation and, consequently, role allocation algorithms can be classified. Subsequently, they established a taxonomy for different task allocation problems. Mainly, they identified three properties to classify multi-robot task allocation:

Single-task robots (ST) vs. multi-task robots (MT): Indicating whether a robot can take on multiple tasks at once (MT) or not (ST). In ALICA, a task allocation problem assumes single-task robots, however due to the hierarchical nature of ALICA programs, and the fact that robots, or teams can execute different plans in parallel, the global problem reflected in ALICA can deal with multi-task robots. Similar to the assumptions made by Gerkey, the decision-relevant entities, such as utilities and preconditions must be independent in the multi-task robot case.

Single-robot tasks (SR) vs. multi-robot tasks (MR): Reflecting whether a task requires exactly one robot (SR) or potentially multiple (MR). Cardinalities in ALICA are used to denote the number of robots each task requires in a specfic context, thereby describing a multi-robot task problem where some tasks and even some robots can be optional, since cardinalities are seen as intervals.

Instantaneous assignment (IA) vs. time-extended assignment (TA):
Indicates whether planned future tasks need to be considered (TA) or that no further information is available regarding potential future tasks (IA). ALICA typically deals with instantaneous assignments due to the dynamics of the environment. However, in principle, the utility functions used by ALICA to evaluate task allocations can incorporate arbitrary information, such as potential future allocations.

Gerkey showed that the ST-SR-IA problem is tractable by relating it to the Optimal Assignment Problem [51]. Furthermore, he showed that the ST-MR-IA problem, which is most relevant to this work, as well as the MT-MR-IA problem are NP-hard. Nair et al. independently showed that the problem of finding an optimal allocation when considering future allocations is NEXP-complete [104].

In their work, they advocated the use of POMDPs (Partially Observable Markov Decision Processes) to model role allocations by introducing role-taking and role-execution actions. Such an approach can lead to optimal decisions if the necessary probability estimates are available and the MDP assumption is applicable. ALICA takes a more general route using utility functions. These can represent decision theoretic entities, such as POMDPS, or simpler heuristic estimates, in case no probability estimates or no proper Markov Model are available.

The problem of dynamic task allocation, where robots continuously need to reevaluate their decisions and reallocate to other tasks in order to adapt to new situations, was studied by Lerman et al. [94]. They advocated a mathematical model to analyse specific multi-robot systems, which allows similar insights as extensive simulations over vast sets of parameters. However, their model is not yet able to deal with real-world robotic teams, which consist of heterogeneous robots whose sensors are subject to noise and which deal with many different environmental information simultaneously.

Naturally, the problem of task allocation was tackled by the RoboCup community as well. Vail and Veloso [170] presented an approach similar to task allocation in ALICA, where robots locally compute the result of a task allocation based on shared information and then act upon it. However, their task allocation algorithm greedily processes an ordered list of tasks and assigns to each task the robot with the highest utility. This process can be embedded into task allocation in ALICA by using an appropriate utility function to order the tasks. Furthermore, in order to stabilise tasks within the team, they specified a minimal time interval between task reallocation. This interval is in the order of several seconds. In ALICA, a threshold can be specified that must be exceeded by a utility difference in order for a reallocation to take place. Additionally, a similarity measure is used to distinguish between reallocations that change little (e.g., exchange the task of two robots) and reallocations that change the complete assignment.

Weigel et al. [178] used utility functions for dynamic role allocation, which measure the utility of each robot-role pair independently of the other robots. Thereby, a robust and efficient role allocation algorithm is created. However, this approach is limited to independent utility measures and to the single-task, single-robot case.

Other coordination and allocation approaches from the RoboCup community often use a central component for some of the decisions to be made. For instance, Lau et al. [93] use a distributed role assignment, but calculate position assignments centrally. For both problems, Lau et al. essentially use a greedy algorithm coupled with a priority list.

Wang et al. [175] suggested another approach, in which robots switch to the least represented role based on a minority game [22]. Thereby, they are able to adapt the role selection strategy dynamically according to the outcome of previous choices. The results are very promising; however, the resulting performance depends on the length of the history considered and the algorithm's space complexity scales exponentially with the history size.

3.6 Estimating Agreement and Conflict Resolution

STEAM allows reasoning about the cost of communication versus the risk to the coordination quality [163], if no communication is used. It then offers repair operators that maintain and fix teamwork as needed. While this approach is appealing, in most real world applications the necessary parameters determining the cost of communication and the risk of failure cannot be precisely determined in advance, as this approach requires. Instead, rough estimates have to be used.

Plan recognition can also be used to detect and thus repair conflicts in teamwork, as done by Kaminka and Tambe [82] through the use of *socially attentive monitoring*. Augmenting our approach with plan recognition is possible and probably beneficial to the overall performance. However, this is out of the scope of this thesis.

The BDI architecture BITE by Kaminka and Frenkel [81] allows for synchronised actions as well as task allocation to be modelled, similar to our approach. BITE offers interchangeable decision protocols for both problems so that a designer can select the best protocol for each specific problem. In a similar fashion, ALICA allows for either weak synchronisation via conditions or more communication intensive strong synchronisation using synchronisation elements (see Section 5.13). Furthermore, ALICA automatically switches decision protocols for task allocation when persistent conflicts are detected (see Chapter 6). BITE does not support automatic switching between protocols. Moreover, in all cases, BITE establishes agreement before acting, while in ALICA actions are started before any communication takes place, thus ALICA has a stronger emphasis on reactivity in dynamic environments. The underlying assumption in ALICA is that the situation might change quicker than messages can be exchanged. Hence, acting immediately is crucial.

Finally, in contrast to all approaches discussed above, ALICA also supports the modelling of team behaviour by constraint systems over real-valued variables and offers a mechanism to coordinate the resulting values in a weak sense. That is, due to the assumed dynamics of the environment, coordination is not geared towards achieving precisely the same instances of vectors in $\mathbb{R}^n$. Instead, coordination establishes a tendency to chose similar values. This method of coordination is discussed in detail in Section 10.3.

3.7 Task Models

The idea of decomposing complex tasks into smaller ones and thereby simplifying the original problem has led to the development of Hierarchical Task Networks (HTN) [140]. HTNs are widely used by planning algorithms, which decompose high-level tasks subsequently into finer-grained structures given action decomposition operators. The result of such a planning step can be expressed as an ALICA plan structure.

The original HTN structure later inspired the TAEMS model (Task Analysis, Environmental Modelling and Simulation) [74]. TAEMS is a very rich multi-agent task model, where each task has a deadline and various different interrelations between tasks can be expressed. Most importantly, TAEMS represents the activity of agents using a distributed goal tree, i.e., in contrast to ALICA, no common program hierarchy is assumed. Instead, agents can potentially discover interrelations to other trees during runtime. In that respect ALICA is more similar to STEAM than to TAEMS due to the assumption of a common program hierarchy. Moreover, instead of the various interrelationships between tasks in TAEMS, ALICA uses finite state machines and supports interrelations between them, i.e., allows transitions to refer to the states inhabited by other agents in other machines. Thus, ALICA takes a more operational stance than TAEMS. Later on, TAEMS was extended with generalized partial global planning (GPGP) [95], yielding a planning oriented approach that is mainly concerned with scheduling concurrent tasks in a distributed environment.

3.8 Constraint-Based Modelling

Chalmers and Gray [23] showed how BDI deliberation can be understood as constraint solving. FLUX [167] represents partial knowledge in the form of partially grounded terms, over which constraints express additional knowledge.

One of the first integrations of constraints in agent specifications was done by Ooi and Ghose [117]. Later it was extended by Dasgupta and Ghose [33], whose approach allows for integration of objectives, which agents try to optimise. Both are based on the BDI language AgentSpeak(L) [129].

They showed that the integration of constraints in high-level agent specification languages can greatly improve efficiency and expressiveness. However, in Ooi and Ghose [117] constraints are only used to select or generate plans based on intentions. We constrain action or behaviour parameters, thereby enabling the specification of behaviour using constraints in a more specific and direct way.

In contrast to these extensions to AgentSpeak(L), our approach models constraints from a global perspective for a team of multiple agents, allows hierarchical decomposition of the constraint problem, such that sub-problems are easier to solve for individual agents, assumes dynamically changing constraints, and integrates coordination into the solving process (see Chapter 8).

De Schutter et al. [38] developed a generic approach to modelling complex spatial tasks for a single robot, which give rise to controller structures to be used during runtime. Their work heavily focuses on control, while we provide generic non-linear constraints, without imposing a control model. In ALICA, the specific controller is to be defined by the executing behaviours.

In [38], constraints were still limited to equality constraints on the controller's output and the joint coordinates. The resulting framework, iTasc, was later extended by Decré et al. [39], making the constraint problem more explicit. The resulting framework allows for problem-specific objective functions and for the constraints to include inequalities. Further, they added non-instantaneous constraint, i.e., constraints over time. However, the problem class remains convex and cannot support arbitrary Boolean combinations of non-linear constraints. The fundamental difference between constraint-based control in iTasc and ALICA is that controllers in iTasc only react to violated constraints while constraints in ALICA are solved before a controller is used. Hence, ALICA potentially spends more time solving than iTasc, but can consider the controlling task on a global scope.

Part II

Propositional ALICA

4 Syntax

In this chapter, the syntactic elements of a propositional variant of ALICA (coined pALICA) are introduced step-by-step. The next chapter introduces corresponding semantics. In Chapter 8, we extend this basic language to the first-order case. Intuitively, the basic problem targeted here can be formulated quite concisely: A team of agents needs to act towards a common goal in a coherent and coordinated manner. In order to reach this goal, the team needs to follow a set of instructions which determine the course of actions. While doing so, they need to continuously monitor their progress in order to detect and avoid potential pitfalls or compensate for problems which have already occurred.

Deriving a set of instructions from a goal description is called planning. Planning, however, is not the focus of this work. Instead, we seek to devise a language in which said set of instructions can be formulated. Furthermore, we are concerned with the efficient and coordinated execution of these instructions.

Imagine for instance two robots that assemble a small structure. A planning algorithm would compute a set of instructions indicating in which order each of them grabs parts, moves, and attaches parts to the structure. The language allows the planner to express the necessary notions such as temporal and causal dependencies between actions. E.g., if robot a assembles part p, robot b might have to wait for this action to complete, so that it can assemble other parts that depend on p. The execution layer of each robot interprets these instructions and executes them. While doing so, it monitors their progress and calls each individual action after the previous action has successfully been completed. Moreover, it reacts to and compensates for failures, such as a robot's unsuccessful attempt to grab a part, or even the complete breakdown of one the robots. Finally, it handles the necessary communication and coordination, such that both robots are aware of the progress of the respective other.

Agents that can act autonomously have a certain set of basic actions at their disposal. The language should allow the identification of the correct action to take for each agent, for each point in time, and under any circumstances. In order to do this, a way of expressing beliefs about the world and conditions on them is needed. We assume the presence of an adequate logic $\mathcal{L}$ with language $\mathcal{L}(Pred, Func)$ for expressing these conditions. There is no requirement posed on this language, though we assume a set of predicates $Pred$ and a set of function symbols $Func$, thus $\mathcal{L}$ has a certain first-order flavour. However, it is entirely possible that $\mathcal{L}$ is compilable into propositional logic.

Furthermore, to describe the specific agents available in a scenario, we refer to the set $\mathcal{A}$, containing all agents that can possibly participate in the team. These two entities, $\mathcal{A}$ and $\mathcal{L}$ make up the domain signature of a pALICA program.

Definition 4.1. A pALICA domain signature $(\mathcal{A}, \mathcal{L})$ consists of

- a set of Agents $\mathcal{A}$, which form the cooperative team;

- a logic $\mathcal{L}$, with language $\mathcal{L}(Pred, Func)$ meant to describe the agents' belief bases with a set of predicates $Pred$ and a set of function symbols $Func$.

By $\vdash_{\mathcal{L}}$ we denote a theorem proving calculus and corresponding algorithm in $\mathcal{L}$, i.e., $\mathcal{F} \vdash_{\mathcal{L}} \phi$ indicates that the algorithm $\vdash_{\mathcal{L}}$ infers the logical consequence ϕ from the set of formulae $\mathcal{F}$. In Chapter 8, we will extend the basic language to incorporate constraint satisfaction problems, and discuss corresponding algorithms in more detail. Here, we assume the presence of a proving algorithm for $\mathcal{L}$. If the logic is clear from the context, we omit it and just write $\vdash$. In propositional ALICA (pALICA), no formula used may have free variables. We denote the set of sentences of $\mathcal{L}(Pred, Func)$ by $\mathcal{L}_S$. Thus, similar to SMT-theories[1], a pALICA program can be interpreted as a propositional formula, whose variables are in turn formulae of another theory, namely $\mathcal{L}$.

In Section 4.1 we introduce the basic action elements available to an agent, *behaviours*. In contrast to classical action calculi, behaviours in ALICA have a significant duration, which typically is only loosely bounded. Section 4.2 continues with higher-level concepts, *plans*, which build a structure based on the atomic behaviours. In the remainder, we discuss additional concepts such as Synchronisations, Roles, and syntactically well-formed programs.

4.1 Behaviours

In real world scenarios, featuring real robots, actions, such as driving to a certain room, can take a significant amount of time, and are non-deterministic; that is, they can fail and even if they succeed, they can have different outcomes. For instance, the action of driving to the kitchen will, when repeated multiple times, almost never bring the executing robot to the exact same spot twice due to noisy sensors and imprecise actuators.

---

[1]   For a discussion of SMT solving, refer to Nieuwenhuis et al. [112].

This brings us to the first elements of the language, namely low-level, atomic activities called *behaviours*, encapsulated in the finite set $\mathcal{B}$. Behaviours usually can only be executed under certain conditions, e.g., it does not make sense to open a door already open, or it is even dangerous to drive onto an unobservable road intersection. Thus, each behaviour has a *precondition* that needs to hold if it is executed:

$$\text{Pre}: \mathcal{B} \mapsto \mathcal{L}_S$$

Hence, $\text{Pre}(b)$ denotes the precondition of behaviour b. Furthermore, since behaviours take a certain timespan to be executed, there may be conditions which need to hold throughout its execution time, called *runtime conditions*:[1]

$$\text{Run}: \mathcal{B} \mapsto \mathcal{L}_S$$

Finally, behaviours are in most cases meant to achieve a certain change in the environment. In other words, an agent executing them tries to achieve a *postcondition*:

$$\text{Post}: \mathcal{B} \mapsto \mathcal{L}_S$$

With these three types of conditions, it is already possible to formulate and solve planning problems, similar to propositional problems formulated in STRIPS [45]. In contrast to action calculi, we do not present a solution for the frame problem [100] and do not define update equations. Instead, we rely on solutions already available (e.g., [92, 134, 150, 168]) and enable the interchangeable usage of ALICA with any belief or knowledge representation and corresponding update semantics.

In this spirit, we enable behaviours to signal successful or unsuccessful termination. In other words, a behaviour — which essentially is a software component implementing one or multiple algorithms, such as controllers or searches of various paradigms — does not need to rely on the implementation of the belief base to encompass the necessary notions nor on the ability of the reasoning algorithm to infer the postcondition. Semantically, this requires two predicates, $\text{Success}(b)$ and $\text{Fail}(b)$ to be represented in $\mathcal{L}(Pred, Func)$, which can be set by behaviour b. Note that these are compilable into propositional logic, as $\mathcal{B}$ is finite.

However, planning algorithms usually rely on the postcondition to find a suitable path to the goal state. In some scenarios, a simulator might instead be employed to execute trial runs, but this is often not feasible. Thus, planning requires the logical equivalence of success signal and postcondition.

---

[1] Runtime conditions relate to invariants in other calculi. However, in contrast to classic invariants, a runtime condition must initially evaluate to true. Furthermore, a runtime condition can change its value, e.g., due to unforeseen changes in the environment, in which case, a corresponding failure is raised.

Definition 4.2. Planning Axiom

$$\Sigma_{plan} \overset{def}{=} \text{Success}(b) \leftrightarrow \text{Post}(b)$$

It is, however, not required of the reasoning algorithm $\vdash_{\mathcal{L}}$ to be able to prove the postcondition in all cases where the behaviour signals success. Though, for any set of beliefs B,

$$B \cup \{\Sigma_{plan} \wedge \text{Success}(b)\} \vdash_{\mathcal{L}} \text{Post}(b)$$

must hold, which essentially comes down to a modus ponens application.

4.2 Plans

Given the atomic behaviours together with their annotations, a structure is needed to formulate more complex recipes or *plans*. This structure constitutes the focus of the language. In general, it describes actions that are to be executed in order to achieve a certain goal, maintain a condition, or control a dynamic system. The simplest way to describe a plan is a list of behaviours to be executed one after the other. There are various representations of such recipes or policies, which extend this simple description.

One prominent representation for action selection is decision-trees, where a decision is made at each node in the tree according to some property of the environment and an action is associated with each leaf-node. Decision-trees and their simpler cousins, decision-lists, have been profoundly researched by the machine learning community. See for instance the works by Quinlan [128] and Murthy [103] on decision-tree learning and [156] for an approach to induction of decision-lists representing single-agent behaviour.

Pure decision-trees can only capture reactive, deterministic behaviour. Probabilistic policies extend that notion to non-deterministic cases, yielding a representation more suitable for classic reinforcement learning techniques. Confronted with more complex tasks, agents need to remember parts of their history. A simple way to capture such a memory in a limited fashion is to use a finite automaton, which can be a much more compact representation than decision-trees. Also note that any decision-tree can be trivially expressed as a finite automaton. More recently, task networks and hierarchical task networks (HTNs) have been used to express agent behaviour, primarily in the context of planning problems [140, 106]. We are aiming for a more expressive language, with additional features to express alternatives, loops, and teamwork aspects such as synchronisations, and dependencies between concurrent tasks.

The basic concept of plans is therefore strongly related to finite automata, which not only allow the formulation of sequences of actions, but also of loops and conditional execution. Hence, every plan p in the set of plans $\mathcal{P}$ is a structure of *states* and the *transitions* between them. The set of all states in a pALICA program is denoted by $\mathcal{Z}$. Transitions intuitively dictate when and how an agent switches from one state to another. Thus, each transition t is a relation between two states and a sentence in $\mathcal{L}$. The transitions of a pALICA program are defined by the set $\mathcal{W} \subseteq \mathcal{Z} \times \mathcal{Z} \times \mathcal{L}_S$. This relationship between transitions and states forms a directed graph for each plan with transitions as edges and states as nodes.

The intuitive idea behind a state is that an agent can, similar to places in petrinets, inhabit a state and while doing so it is obliged to do something, i.e., execute one or multiple behaviours. The function Behaviours: $\mathcal{Z} \mapsto 2^{\mathcal{B}}$ defines which behaviours are contained in a state and are therefore to be executed by inhabiting agents.

A finite automaton has an initial state, indicating where to start a calculation. A single initial state, however, does not suffice to grasp the behaviour of multi-agent systems. Instead, we allow for multiple initial states, each one defining a starting point for a sub-group of agents. The automaton is thereby divided into sub-automata, which are not necessarily connected. This way, it is possible to describe activities for multiple agents which depend on each other. Section 5.13 introduces a method for describing such dependencies and thereby coordinating different activities within a plan. For now, we introduce the notion of a *task*, which points to an initial state within a plan, and thus identifies a sub-automaton within it. The set of all tasks of a program is referred to as $\mathcal{T}$. Note that while we do not enforce the sub-automata identified by each task to be disjoint, disjoint sub-automata are beneficial, since in this case each state unambiguously identifies a task. This can be exploited to limit communication overhead (see Section 7.5).

The partial function Init: $\mathcal{P} \times \mathcal{T} \mapsto \mathcal{Z}$ maps a task and a plan to an initial state, e.g., Init(p, τ) is the initial state of task τ in plan p. This allows for tasks to be independent of a specific plan, a feature useful later on when we consider sets of plans. Given tasks as means to identify different parts of a plan, agents can be assigned to and work on different parts of the plan. However, some tasks might still require multiple agents. An object that needs to be moved might be too heavy for a single agent or an area too large to be searched in feasible time. Hence, we allow multiple agents to execute the same task within a plan. The number of agents executing a task might be subject to some constraints. We capture such bounds on the number of agents by the partial function $\xi : \mathcal{P} \times \mathcal{T} \mapsto \mathbb{N}_0 \times (\mathbb{N}_0 \cup \{\infty\})$, such that $\xi(p, \tau) = (n_1, n_2)$ indicates how many agents must at least (n_1) and may at most (n_2) execute task τ in plan p.

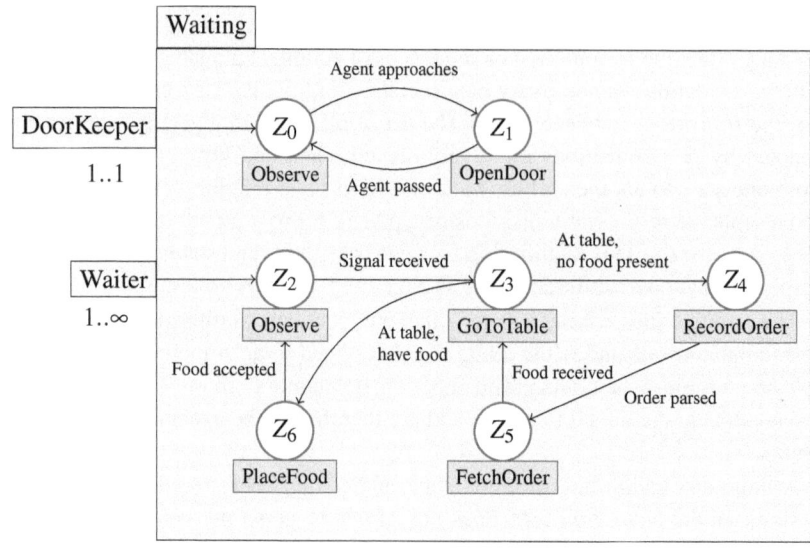

Figure 4.1: Example Plan: Waiting

Some plans are meant to maintain certain conditions and run indefinitely. Others, however, are goal-directed and terminate once the intended environmental state is reached. Reflecting this, we incorporate terminal states into the plan structure. Let States(p) denote the states of plan p, then some states Success$(p) \subseteq$ States(p) are successful terminal states. Once an agent has reached such a state, it has successfully completed its task. Similarly, by Fail$(p) \subseteq$ States(p), we denote the set of terminal states that indicate a failure. An agent reaching a failure state has failed to complete its task. In Section 5.5, we will discuss successful termination in more detail. Section 5.13.2 presents possibilities to deal with failures.

With the elements defined so far, we can already express some complex behaviour, such as the example in Figure 4.1 from the service robotic domain. The plan *Waiting* contains two tasks: one for a set of robots serving the tables and one for a single robot taking care of the door to the service area.

However, there are two major shortcomings. Firstly, as problem domains become complex, so do plans that specify solutions and we have not yet introduced any way to abstract complexity. Secondly, it is cumbersome to express alternatives in a single state machine. While alternative transitions are possible, these quickly converge towards fully connected state machines if the scenario requires quick adaptation. Both problems require ways to compare plans with each other.

Hence, we introduce conditions for plans similar to those defined for behaviours. We extend the functions Pre and Run to map the set of plans to sentences of $\mathcal{L}$ as well:

$$\text{Pre}\colon \mathcal{P} \cup \mathcal{B} \mapsto \mathcal{L}_S$$
$$\text{Run}\colon \mathcal{P} \cup \mathcal{B} \mapsto \mathcal{L}_S$$

Facilitating planning, a plan should also feature a postcondition. Since we defined the notion of terminal states, it is sensible to attach postconditions to successful terminal states. This allows for different, yet successful outcomes to be modelled for each plan. Similarly, failure states are also equipped with a postcondition. This allows for failure analysis by a reasoning component. We extend Post to the partial function

$$\text{Post}\colon \mathcal{Z} \cup \mathcal{B} \mapsto \mathcal{L}_S$$

which maps terminal states to postconditions. On the semantic level, there is a major difference between the postcondition of a state and that of a behaviour. If a terminal state is reached, this indicates that the postcondition holds. However, a postcondition being evaluated to true given the internal belief of an agent does not implicate that an agent, or even the corresponding agent inhabits the corresponding terminal state. When introducing runtime semantics in Section 5.13, we will come back to this distinction.

With the set of conditions introduced so far, plans can be evaluated with respect to a situation, be it the current one or a hypothesised situation during planning. Assuming $\mathcal{L}(Pred, Func)$ is monotonic[1], it is difficult to express preferences for one plan over others should multiple pre- and runtime conditions hold.

Therefore, we introduce a fourth element to evaluate a plan: *utility functions*. A utility function maps a situation, described by a belief base, i.e., by a set of formulae in $\mathcal{L}(Pred, Func)$ to a set over which a total order is defined. Following a popular choice, we assume utility functions map onto the real numbers.

$$\mathcal{U}\colon \mathcal{P} \mapsto 2^{\mathcal{L}_S} \mapsto \mathbb{R}$$

Each plan $p \in \mathcal{P}$ has a utility function, denoted by $\mathcal{U}(p)$, which in turn maps a set of formula to the real numbers. Thus, it rates a plan according to a situation. Due to the fact that plans can have multiple initial states, it can also rate different ways of executing plans, i.e., different *allocations*. This notion will be discussed in Section 5.8.

---

[1]   See [14] for how preferences relate to non-monotonic logics.

Given utility functions and pre- and runtime conditions, plans are comparable with respect to a situation. To allow agents to select a plan from a set of possible alternatives through comparison, we group plans together in what we call *plantypes*. Such grouped plans can be (but are not necessarily) directed towards the same goal. The set of all plantypes in a pALICA program is referred to as $\mathcal{P}_\vee \subseteq 2^{\mathcal{P}}$. A plan p belonging to plantype P is also called a *realisation* of P.

Fostering the idea of hierarchies, we introduce these sets as properties of states. Thus, each state can contain a number of sets of alternative plans in the same manner as it can contain behaviours. The function

$$\text{PlanTypes}: \mathcal{Z} \mapsto 2^{\mathcal{P}_\vee}$$

maps each state onto a possibly empty set of sets of alternative plans such that each plantype is meant to be executed in parallel.

The relationships between states and plantypes and between plantypes and plans span a directed graph between plans. In Section 4.5, we will constrain this graph to a tree. The top-most element of this tree is identified by the plan p_0. It contains a single task τ_0, and a single state z_0. Thus, p_0 can be seen as the main procedure of the resulting multi-agent program. We refer to this structure as *plan-tree*.

Figure 4.2 shows how the plan *Waiting* in Figure 4.1 can be simplified by hierarchical abstraction. Two new plantypes are introduced: one for handling single tables and one for fetching food from the service room. Both abstract away the complexity of each task. This also fosters the possibility of reuse. For instance, a plan for fetching food from the service room can now be used in other settings. The plans *GetOrder* and *BringFood* are two realisations of *HandleTable*, each handling a specific case, distinguished by their respective preconditions. Note that only *BringFood* ends in a terminal state which indicates the success of handling the specific table.

The plan structure as introduced so far will play a major role in the next chapters. Therefore, we introduce some macros simplifying expressions over plans:

- By $\text{PlanTypes}^+(z)$, we denote a transitive flavour of PlanTypes, inductively defined by:

  - $\text{PlanTypes}(z) \subseteq \text{PlanTypes}^+(z)$

  - If $P \in \text{PlanTypes}^+(z)$ then
    $(\forall z', p) p \in P \wedge z' \in \text{States}(p) \rightarrow \text{PlanTypes}(z') \subseteq \text{PlanTypes}^+(z).$

- Plans: $\mathcal{Z} \mapsto 2^{\mathcal{P}}$, $\text{Plans}(z)$ denotes the set of plans that can be executed within state z:

$$\text{Plans}(z) \overset{def}{=} \bigcup \text{PlanTypes}(z)$$

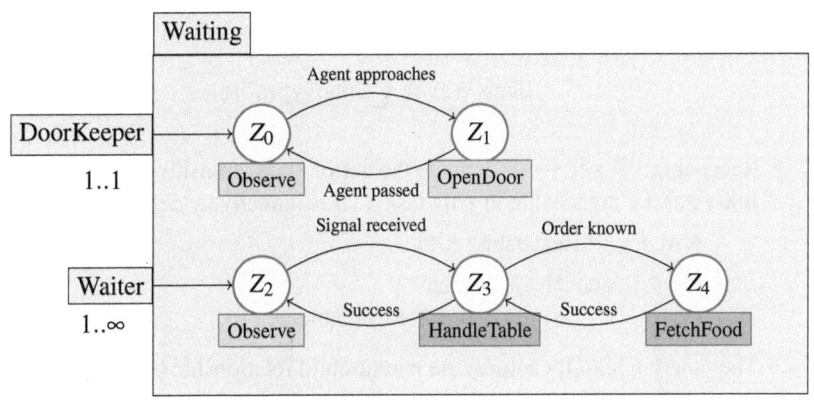

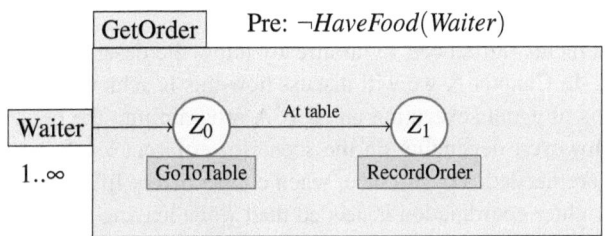

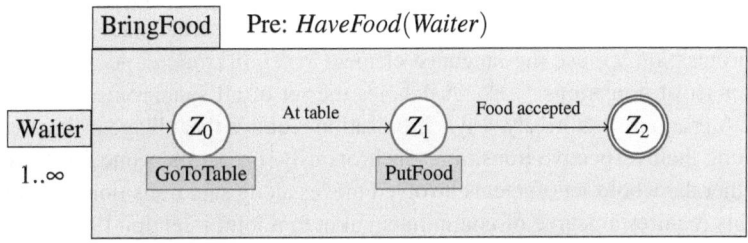

Figure 4.2: Example Plan: Hierarchical Waiting

- Plans$^+$: $\mathcal{Z} \mapsto 2^{\mathcal{P}}$ is the transitive closure of Plans, defined by:

$$\text{Plans}^+(z) \stackrel{def}{=} \bigcup \text{PlanTypes}^+(z)$$

- Reachable: $\mathcal{P} \times \mathcal{T} \mapsto \mathcal{Z}$ denotes the set of states transitively connected to Init(p, τ), i.e., reachable in p by task τ. It is inductively defined by:
  - Init$(p, \tau) \in$ Reachable(p, τ)
  - If $z \in$ Reachable(p, τ) then
    $(\forall z', \phi)(z, z', \phi) \in \mathcal{W} \rightarrow z' \in$ Reachable(p, τ)

- The macro ChildOf captures the parent-child relationship between plans:

$$\text{ChildOf}(p, p') \stackrel{def}{=} (\exists z)z \in \text{States}(p) \wedge (\exists P)P \in \text{PlanTypes}(z) \wedge p' \in P$$

The transitive closure of ChildOf is denoted by ChildOf$^+$.

4.3 Synchronisations

The language elements introduced so far already allow the description of coordinated behaviour. In Chapter 5, we will discuss how this is achieved by reflecting the internal status of agents executing an ALICA program into the belief base of other agents. However, depending on the scenario, different levels or strengths of coordination are needed. For instance, when cooperatively lifting a potentially fragile object a tighter coordination is needed than when leaving a room one after the other. In the limit, actions might need to be synchronised as tightly as communication latencies allow.

We therefore introduce an additional language element, following the idea of synchronisations formulated by Kinny et al. [85]. In contrast to their work, we use an explicit language element to distinguish between loosely synchronised and tightly synchronised execution. Loose synchronisation is achieved simply by referring to the local representation of other agents' internal state. For tighter synchronisation we use the language element *synchronisation*. A synchronisation s is a set of transitions $\subseteq \mathcal{W}$. Λ denotes the set of all synchronisations in an ALICA program. Intuitively, a synchronisation requires that all agents involved move along their respective transition synchronously, i.e., at the same time. Moreover, either the whole set of agents involved moves along said transitions, or none at all. This requires a degree of commitment akin to a joint intention [97]. Similar constructs have also been introduced in other languages, such as synchronising speech acts in the planning language MAPL [13].

4.4 Roles

Plans describe activities, i.e., they describe how a certain goal can be achieved. Tasks denote specific parts within plans. In order to assign agents to tasks, a third concept is needed, which we refer to as *roles*. Roles are assigned to agents, allowing to evaluate an agent's adequateness for a certain task and establish a general team make-up. Following Wooldridge et al. [182], roles in ALICA reflect an agent's expected function within the team. This functionality is determined by the problem domain, the capabilities of the agent, and the team composition. For instance, an agent equipped with highly sophisticated sensors can perform searches more efficiently than an agent with moderate sensing capability. This leads to a definition of roles and tasks as described by Campbell and Wu [20].

Let $\mathcal{R}$ denote the set of all roles of a pALICA program. Then the function Pref: $\mathcal{R} \times \mathcal{T} \mapsto [-1,1]$ maps a role and a task to a real number in the interval $[-1,1]$, expressing the *preference* of tasks towards specific roles.

There are various approaches to the problem of assigning roles to agents. In the simplest case, each agent is assigned a single static role. The resulting problem has been described by Gerkey [55] as the single-task, single-robot instantaneous assignment (ST-SR-IA) and can be solved in polynomial time.

More complex techniques allow for adaptation in case a team member breaks down or is partially incapacitated and allow multiple roles per agent, multiple agents per role, and possibly consider future events. In this case, the complexity of the problem increases to NP [55] or even to NEXP-time if possible future reallocations are considered [104]. Possible role assignment algorithms were recently discussed by Campbell and Wu [21].

For the execution of an ALICA program, it is required that each agent is assigned a non-empty set of roles and that this set is known throughout the team. We do not limit ALICA to a specific role allocation algorithm and thus only give a general description of the necessary concepts in ALICA. In Section 5.6, we introduce a possible instance of a suitable role allocation algorithm.

In order to achieve role allocations according to the different skills and abilities of the agents involved, another concept is needed: *Capabilities*. A capability is a description of a specific skill, or ability, together with a degree rating it. Informal sentences such as "Robot r is very fast." or "This role requires grasping." refer to capabilities. Again, we do not pose any restrictions on how capabilities are represented here, but discuss them further in Section 5.6, where we introduce a concrete role allocation algorithm. By $\mathcal{C}$, we denote the set of all capabilities.

The function Cap: $\mathcal{R} \cup \mathcal{A} \mapsto 2^{\mathcal{C}}$ maps agents onto their *provided* capabilities and roles onto their *required* capabilities. Given agents, roles, and their respective required and provided capabilities, a role allocation algorithm can match agents to roles. However, this can only be done unambiguously if there are as many roles as agents and each agent must take on exactly one role (ST-SR). Furthermore, additional information cannot be taken into account. Hence, we introduce a third concept: *Formations*. A formation fills the gap by providing information about which roles are essential and which are prioritised over others given a specific set of agents and potentially a situation in which to decide. As the concept of formations highly depends on the role allocation semantics, we leave it open and discuss an example in Section 5.6.

4.5 Well-Formedness

In the previous sections, we introduced the language elements of pALICA and stated some relationships between them, such as the plan-tree. We now constrain the syntax of a pALICA program in a syntactical manner, in order to guarantee the intended structure of these relationships. This leads to the notion of well-formed pALICA programs. Let Σ_{syn} be the set containing the following axioms:

- The top-level plan contains precisely one state, z_0, and one task, τ_0:
$$\text{States}(p_0) = \{z_0\} \wedge \text{Tasks}(p_0) = \{\tau_0\} \tag{4.1}$$

- States belong to at most one plan:
$$(\forall p, p' \in \mathcal{P}) \, \text{States}(p) \cap \text{States}(p') = \emptyset \vee p = p' \tag{4.2}$$

- No transition connects states in different plans:
$$(\forall(z_1, z_2, \phi) \in \mathcal{W})(\exists p \in \mathcal{P})z_1 \in \text{States}(p) \wedge z_2 \in \text{States}(p) \tag{4.3}$$

- Synchronisations happen only within a plan:
$$(\forall s \in \Lambda)(\forall w, w' \in s)(\exists z_1, z_2, z_3, z_4, \phi_1, \phi_2)w = (z_1, z_2, \phi_1) \tag{4.4}$$
$$\wedge w' = (z_3, z_4, \phi_2) \wedge (\exists p \in \mathcal{P})z_1 \in \text{States}(p) \wedge z_3 \in \text{States}(p)$$

- Failure and success sets are disjoint subsets of the corresponding state set:
$$(\forall p \in \mathcal{P}) \, \text{Success}(p) \cup \text{Fail}(p) \subseteq \text{States}(p) \tag{4.5}$$
$$\wedge \, \text{Success}(p) \cap \text{Fail}(p) = \emptyset$$

- A task associated with a plan identifies an initial state within that plan:
$$(\forall \tau \in \text{Tasks}(p))(\exists z \in \text{States}(p)) \, \text{Init}(p, \tau) = z \tag{4.6}$$

- All plan-task pairs have a valid cardinality interval associated:
$$(\forall p \in \mathcal{P}, \tau \in \mathcal{T})\tau \in \text{Tasks}(p) \rightarrow \tag{4.7}$$
$$(\exists n_1, n_2) \, \xi(p, \tau) = (n_1, n_2) \wedge n_1 \leq n_2$$

- There is a postcondition associated with each success and failure state:

$$(\forall p \in \mathcal{P})(\forall z \in \mathrm{Success}(p) \cup \mathrm{Fail}(p))(\exists \phi)\,\mathrm{Post}(z) = \phi \qquad (4.8)$$

- Terminal states do not have sub-plans or behaviours attached:

$$(\forall z)((\exists p)z \in \mathrm{Success}(p) \vee z \in \mathrm{Fail}(p)) \rightarrow \mathrm{PlanTypes}(z) = \emptyset \qquad (4.9)$$
$$\wedge \mathrm{Behaviours}(z) = \emptyset$$

- The top-level plan is connected to all plans:

$$(\forall p)p = p_0 \vee \mathrm{ChildOf}^+(p_0, p) \qquad (4.10)$$

- Every plan has at most one parent:

$$(\forall p, p_1, p_2)\,\mathrm{ChildOf}(p_1, p) \wedge \mathrm{ChildOf}(p_2, p) \rightarrow p_1 = p_2 \qquad (4.11)$$

- The transitive closure of the parent-child relationship is asymmetric:

$$(\forall p, p')\,\mathrm{ChildOf}^+(p, p') \rightarrow \neg\,\mathrm{ChildOf}^+(p', p) \qquad (4.12)$$

An ALICA program is well-formed if and only if it satisfies Σ_{syn}.

Axiom 4.1 guarantees the existence of a unique root node in the plan-tree. This simplifies execution and coordination, as the number of possible states any agent can have with respect to the highest level is limited: either the agent participates or it does not. The axioms 4.2 to 4.6 limit the internal structure of plans to be local, that is, elements belonging to different plans are not connected. Axioms 4.6 to 4.8 guarantee the existence of some elements essential for execution, such as cardinalities for task-plan tuples. Note that trivial elements such as the cardinality $(0, \infty)$ and the postcondition $\top$ are allowed.

Axiom 4.9 disallows terminal states to have children. Although this is not strictly necessary, terminal states intuitively terminate execution and should not contain internal programs.

The axioms 4.10 to 4.12 constrain the plan-tree structure to a finite tree. More specifically, Axiom 4.10 enforces that the top-level plan p_0 is transitively connected to all plans. Axiom 4.11 enforces that each plan has at most one parent. Together with Axiom 4.10, this entails that every plan other than p_0 has exactly one parent. Finally, Axiom 4.12 enforces asymmetry of the transitive closure of the parent-child relationship. Since $\mathcal{P}$ is finite, the plan-tree is also finite. Note that p_0 cannot have a parent due to Axiom 4.10 and Axiom 4.12.

From a software architectural view, enforcing a tree-structure seems to be overly restrictive, and indeed on the implementation level, a directed acyclic graph is supported, allowing for plans to be reused in different contexts. The runtime engine recognises the different plan instances and handles them accordingly as if they were different plans. We constrain the structure to a tree in order to avoid overly complex notations, which would be required when distinguishing plans and plan instances. Note that a directed acyclic graph can trivially be transformed into a tree by duplicating nodes with multiple parents, so the expressiveness is not restricted with respect to acyclic graphs.

However, for both theory and implementation, the plan-tree must not contain cycles. While permitting cycles would allow for recursive definitions and thus increase the expressiveness, the semantics of recursion in the presence of multiple agents are not at all straight-forward. For instance, each agent could be operating at different recursion depths at a given time. We therefore exclude recursion.

4.6 Overview of the Syntactic Elements in pALICA

Summarising, a pALICA program contains only very few different elements. Firstly, the domain signature defines the set of possible agent, as well as the logic expressing beliefs about the world. The central element of pALICA are plans, which formulate recipes to tackle given problems. Plans consist of states and transitions between them. Transitions can be linked together by synchronisations in order to enforce a tight coupling between agents. States can contain plantypes, which represent sets of alternative plans. Moreover, states can contain behaviours, which are atomic action programs. This relationship yields a tree-shaped graph whose nodes are plans. Initial states within plans are tagged by tasks to which agents can be allocated. Such an allocation takes the roles of each agent into account. Roles are assigned to agents given their respective required and provided capabilities.

Table 4.1 summarises the formal elements of pALICA. With the exception of $\mathcal{L}(Pred, Func)$, all the sets are finite. This allows for compilation into propositional logic, if $\mathcal{L}(Pred, Func)$ is compilable into propositional logic. The individual elements are structured using the (partial) functions listed in Table 4.2.

$(\mathcal{A},\mathcal{L})$	the domain signature	The domain signature consists of the set of possibly interacting agents and the logic with which the world is represented.
$\mathcal{R}$	a set of roles	This set contains all availables roles any agent can be assigned to.
$\mathcal{B}$	a set of behaviours	Behaviours are atomic action programs which form the means to interact with the environment.
$\mathcal{P}$	a set of plans	Each plan describes a specific cooperative activity.
$\mathcal{P}_\vee$	a set of plantypes	A plantype is a set of alternative plans.
$\mathcal{T}$	a set of tasks	Each task intuitively describes a function or duty within plans, meant to be fulfilled by one or more agents.
$\mathcal{Z}$	a set of states	A state occurs within a plan as a step during an activity. It can contain plantypes and behaviours.
$\mathcal{W}$	a set of transitions	Each transition (z_1, z_2, ϕ) relates a predecessor state z_1 with a successor state z_2 and a condition $\phi \in \mathcal{L}(Pred, Func)$.
Λ	a set of synchronisations	Each relating transitions in order to express the need for synchronised actions.
$\mathcal{C}$	the set of all capabilities	Capabilities are used to match agents and roles.
p_0, τ_0, z_0	top-level elements	The top-level plan, task, and state respectively form the root node of the program graph.

Table 4.1: Elements of a pALICA Program

States: $\mathcal{P} \mapsto 2^{\mathcal{Z}}$	States maps plans to the set of contained states.
Tasks: $\mathcal{P} \mapsto 2^{\mathcal{T}}$	Tasks maps plans to the set of related tasks.
$\xi: \mathcal{P} \times \mathcal{T} \mapsto \mathbb{N}_0 \times (\mathbb{N}_0 \cup \{\infty\})$	ξ defines the upper and lower bound of agents assignable to a task τ in plan p.
Init: $\mathcal{P} \times \mathcal{T} \mapsto \mathcal{Z}$	Init maps a plan and a task to the corresponding initial state.
Pre: $\mathcal{P} \cup \mathcal{B} \mapsto \mathcal{L}_S$	Pre(p) denotes the precondition of plan or behaviour p.
Run: $\mathcal{P} \cup \mathcal{B} \mapsto \mathcal{L}_S$	Run(p) denotes the runtime condition of plan or behaviour p.
PlanTypes: $\mathcal{Z} \mapsto 2^{\mathcal{P}_\vee}$	PlanTypes(z) denotes the set of plantypes to be executed in state z.
Behaviours: $\mathcal{Z} \mapsto 2^{\mathcal{B}}$	Behaviours(z) denotes the set of behaviours to be executed in state z.
Success: $\mathcal{P} \mapsto 2^{\mathcal{Z}}$	Success(p) denotes the set of terminal states of plan p, which indicate successful execution of the plan.
Fail: $\mathcal{P} \mapsto 2^{\mathcal{Z}}$	Fail(p) denotes the set of terminal states of plan p, which indicate unsuccessful execution of the plan.
Post: $\mathcal{Z} \mapsto \mathcal{L}_S$	Post(z) is a partial function, that maps terminal states of a plan to postconditions.
$\mathcal{U}: \mathcal{P} \mapsto 2^{\mathcal{L}_S} \mapsto \mathbb{R}$	$\mathcal{U}(p)$ is the utility function of p, evaluating p with respect to a set of formula.
Pref: $\mathcal{R} \times \mathcal{T} \mapsto [-1, 1]$	Pref(r, τ) is the preference of task τ towards role r.

Table 4.2: Structure Definitions of a pALICA Program

5 Semantics

In the previous chapter, we introduced the language pALICA and its syntax. We thereby distinguished between syntactically incorrect and well-formed programs. Here, we will introduce a meaning to the latter.

The semantics defined here serve three purposes. Firstly, it should yield an intuitive understanding of pALICA programs, such that a developer can easily design multi-agent programs. Secondly, it specifies in detail how agents execute a program and thus defines a range of possible execution engines. Thirdly, it should allow mechanical proofs of various properties, such as liveliness or safety conditions. Out of the three, we regard the second purpose as the most important, since we are mostly interested in practical applications, and thus, in the execution of programs.

In the following, we declare a set of principles the semantics should follow. Afterwards, we begin introducing the semantics with an agent model that defines in a most general way how agents are understood in the context of ALICA. Afterwards, we define operational semantics in terms of a transition system [122] which dictates how the internal representation of an agent changes over time given a specific ALICA program.

5.1 Fundamental Principles

The design of the ALICA semantics is guided by the following principles: Domain independence, autonomy, and locality. They are meant to guarantee portability, scalability, and robustness against adverse domain features such as unreliable communication.

Domain Independence Although robotic scenarios are in the focus of this work, ALICA is designed domain independently, such that the results of this work can be used in other domains than those discussed here. Therefore, ALICA makes as little assumptions as possible about how an agent represents its environment. For instance, it is entirely possible to use a closed world assumption or an open world assumption when referring to the environment. Moreover, one can use a classical first-order logic, a modal logic, or even a hybrid approach such as a probabilistic logic. A specific ALICA program references the world representation through the language of $\mathcal{L}$, which must be defined as part of the signature.

Autonomy   One of the requirements we formulated in Chapter 1, is the ability to cope with unreliable communication. ALICA tackles this problem by performing calculations redundantly, that is, decisions regarding the team are made by all agents individually and autonomously. Inconsistent decisions can be subsequently detected and corrected once corresponding information is available, e.g., through communication or action recognition. This principle allows ALICA to operate even under highly degraded network conditions, as shown in [159]. Due to redundant calculations, message delivery times can be almost unbounded. There is only one exception to this rule, which is related to the recognition of an incapacitated agent. Our implementation assumes that if no message was received from an agent for a certain period of time, this agent is no longer able to function properly. The time period used depends on the expected network quality, the degree of dynamics in the domain, and the likelihood of an agent breaking down. In Section 7.5, we will discuss this in more detail.

Locality   The global state of a team of agents executing an ALICA program is represented by the combined states of all agents involved. In order to cope with the potential complexity of the problem tackled by the team, ALICA exploits the hierarchical structure of the program. Solutions for encountered problems are described by plans, which split the team using tasks, and employ sub-plans to solve potential sub-problems.

Maintaining this tree structure and performing all necessary calculations to make all decisions involved can very well overwhelm the computational power of a single agent, which might be why a multi-agent system was chosen in the first place. Hence, ALICA adopts a *locality* principle. Each agent keeps track of the plans it is involved in and only participates in decisions regarding these plans. Parts of the global state of the team in which an agent does not participate are ignored by this agent.

If the problem can be decomposed into sub-problems, this principle significantly reduces the computation costs in large teams. Thereby the locality principle fosters scalability with respect to the number of agents participating. This is also a principle that comes very naturally to us human beings. For instance, during soccer, an attacker currently dribbling towards the opponent goal, will not concern herself with the precise positioning of her team members defending the goal, she is content knowing that they are defending. This is equivalent to the ALICA notion of knowing which agents are allocated to which task at a higher level of the plan hierarchy. The same principle is also applied by humans in less dynamic scenarios. If your department participates in a large collaborative project, you will not concern yourself with every detail of your partners' doings. Instead, it suffices to know an abstracted progress status.

5.2 Agent Model

In Section 2.1, we introduced the concept of agents. Following the most general model of an intelligent agent described there, we do not impose any restrictions, but embed the ALICA control structure in the appropriate place. Assuming a control loop consisting of the three parts *Belief Update*, *Reasoning*, and *Execution*, ALICA integrates itself into the reasoning step.

The preceding belief update incorporates perceived information in the agent's internal model of its surroundings. Within an ALICA program, $\mathcal{L}(Pred, Func)$ can be used to express facts and relationships within this model. The reasoning step involves any kind of internal calculations, such as planning and intention update. Most importantly, the internal state of the executed ALICA program is updated in this step. In other words, an ALICA computation step is performed.

The execution phase executes any action command as decided by the reasoning steps. In the context of ALICA, this refers to the atomic behaviour programs which issue action commands. These commands can be processed further by intermediate reasoning steps. Finally, this last step also includes any active communication, as the act of sending information is regarded as an action by itself.

We do not require agents to follow the model in Figure 2.1 in a strict sense. It is merely a guideline of necessary steps, and their suggested order. Especially given today's parallel hardware architectures, a more modular architecture is sensible. However, it still holds that perceptions need to be integrated into the internal model, the result should be reacted upon in reasoning steps, and afterwards necessary action updates should be made. In Chapter 7, we will introduce a more elaborate software architecture.

From the point of view of ALICA, an agent has a *configuration* at any point in time. This configuration reflects its internal status, comprising of its belief base, its roles, and its state with respect to the ALICA program in execution.

Definition 5.1 (Agent Configuration). For any agent $a \in \mathcal{A}$, let $\mathrm{Conf}(a)$ denote its *configuration*. An agent configuration in pALICA is a tuple $(B, \Upsilon, E, \mathrm{R})$, where

- B is the agent's belief base.

- Υ is the agent's plan base.

- $E \subseteq \mathcal{B} \times \mathcal{Z}$ is the agent's execution set, i.e., the set of behaviour-state tuples the agent executes.

- R is the set of roles a currently holds.

The belief base holds the current model of the environment – everything the agent believes to be true. The plan base is a representation of the agent's current state within the program. The execution set contains all tuples (b, z) of behaviours b the agent executes together with the state z in which b occurs, called the *context* of b. R contains all roles a holds within the team.

Initially, an agent's configuration is empty except for domain-specific beliefs. Hence, after start up, an agent does not make any assumptions about the current state of the team. In particular, it does not believe to be executing any plan.

Definition 5.2. An initial agent configuration has the form $(B, \emptyset, \emptyset, \emptyset)$. Such that B does not contain any belief referring to the status of the ALICA program.

The operational semantics define how agent configurations are updated as the agent executes a program. Before we introduce corresponding rules, we will discuss the elements of agent configurations in more detail.

5.2.1 Plan Base

The plan base of an agent captures its current intentional state. Specifically, it denotes which state the agent inhabits for each plan it participates in and which tasks it committed to. Hence, each element holds a procedurally represented intention.

Definition 5.3 (Plan Base). An agent's plan base is a set of triples (p, τ, z), consisting of a plan p, a task τ, and a state z. The plan base of an agent a is denoted by $\mathrm{PBase}(a)$.

If (p, τ, z) is an element of $\mathrm{PBase}(a)$ for an agent a, we say a participates in p (or executes p), is committed to task τ and inhabits state z. An agent cannot arbitrarily participate in plans, but instead can only take on one task per plan at a time, and only execute one plan per plantype. Furthermore, it can only commit to tasks belonging to the respective plan and inhabit only states reachable within that task. This restriction is expressed by the *plan base axioms*.

Definition 5.4. The set of plan base axioms Σ_p contains exactly the following:

$$(p, \tau, z) \in \text{PBase}(a) \wedge (p, \tau', z') \in \text{PBase}(a) \rightarrow \tau = \tau' \wedge z = z' \tag{5.1}$$

$$(p, \tau, z) \in \text{PBase}(a) \wedge (p', \tau', z') \in \text{PBase}(a) \wedge p \in P \wedge p' \in P \rightarrow p = p' \tag{5.2}$$

$$(p, \tau, z) \in \text{PBase}(a) \rightarrow \tau \in \text{Tasks}(p) \wedge z \in \text{States}(p) \wedge z \in \text{Reachable}(p, \tau) \tag{5.3}$$

We define the following macro over plan bases, which captures their hierarchical structure:

Definition 5.5. $\text{Plans}^+(\Upsilon, z)$ denotes the set of plans that are executed by an agent with plan base Υ in the context of z. It is defined inductively as the smallest set such that:

- $p \in \text{Plans}^+(\Upsilon, z) \leftarrow (p, \tau, z') \in \Upsilon \wedge p \in \text{Plans}(z)$

- $p \in \text{Plans}^+(\Upsilon, z) \leftarrow p' \in \text{Plans}^+(\Upsilon, z) \wedge (p', \tau', z') \in \Upsilon \wedge p \in \text{Plans}(z') \wedge (p, \tau, z'') \in \Upsilon$

On the implementation level, a plan base might be represented using a structure other than a set. Using a graph structure allows for swifter execution of the rule set. Depending on the rule, a single instance can be executed in $O(1)$ or $O(n)$ instead of $O(n^2)$, where n is the number of triples in the plan base. In Section 5.14, we will show that if the pALICA program is well-formed, then the plan base always forms a tree. This property further simplifies the runtime representation of the program. Here, we assume a set representation, which allows for a more intuitive description of the semantics, rather than focussing on efficiency. In Chapter 7, we will discuss implications for the implementation in more detail.

5.2.2 Belief Base

The belief base represents the internal model an agent has of the environment, which includes its beliefs about cooperative agents executing the same plan. Within ALICA, conditions and utilities refer to this representation using $\mathcal{L}(Pred, Func)$. Thus, ALICA makes some assumptions concerning the representation of beliefs about team members.

Let $\mathcal{L}(Pred, Func)$ be the language of the belief base. $\mathcal{L}$ is a first-order logic[1] extended by the modal operators Bel_a and K_a, for each agent a in $\mathcal{A}$.

---

[1]   In principle, ALICA can also work with a propositional modal logic, since the set of ground terms is finite.

The operator Bel_a expresses individual belief of agent a. $\text{Bel}_a \phi$ denotes that agent a believes ϕ. Formally and according to the knowledge axioms by Fagin et al. [42], Bel is defined (as KD45 system) by the following axioms:

- $\text{Bel}_a(\phi \to \psi) \to (\text{Bel}_a \phi \to \text{Bel}_a \psi)$ (Distribution Axiom)
- $\text{Bel}_a \phi \to \neg \text{Bel}_a \neg \phi$ (Consistency Axiom)
- $\text{Bel}_a \phi \to \text{Bel}_a \text{Bel}_a \phi$ (Positive Introspection Axiom)
- $\neg \text{Bel}_a \phi \to \text{Bel}_a \neg \text{Bel}_a \phi$ (Negative Introspection Axiom)
- $(\forall x) \text{Bel}_a \phi(x) \to \text{Bel}_a((\forall x)\phi(x))$ (Knowledge Quantifier)

These axioms form the notion of strongly rational belief. The modality K extends Bel towards knowledge: $K_a \phi \overset{def}{=} (\text{Bel}_a \phi) \wedge \phi$.

Building on individual belief, we use the usual notions of "everyone believes", EBel, and mutual belief, MBel. Everyone believes is defined by (after Rao et al. [130]):

$$\text{EBel}_A \phi \overset{def}{=} (\forall a \in A) \text{Bel}_a \phi$$

The formula $\text{EBel}_A \phi$ is satisfied if and only if all agents a in group A believe ϕ. The mutual belief of ϕ is defined as all agents a of a group A believing ϕ and all of them believing this mutually. More formally, $\text{MBel}_A \phi$ is defined as the greatest fixpoint of:

$$\text{MBel}_A \phi = \text{EBel}_A \phi \wedge \text{EBel}_A \text{MBel}_A \phi$$

This yields an infinite conjunction of the form:

$$\text{MBel}_A \phi \leftrightarrow \text{EBel}_A \phi \wedge \text{EBel}_A \text{EBel}_A \phi \wedge \ldots \wedge \text{EBel}_A \ldots \text{EBel}_A \phi \wedge \ldots$$

In words, everyone believes ϕ, believes that everyone believes ϕ, believes that everyone believes that everyone believes ϕ, and so on.

The set of terms in $\mathcal{L}(Pred, Func)$ contains certain ALICA specific terms, such as a constant for each agent in $\mathcal{A}$, and domain-specific terms, such as coordinates or terms representing physical objects. Formally, the set of terms in $\mathcal{L}(Pred, Func)$ is given by:

- A countable infinite set of variables, $X = \{x_1, x_2, \ldots, x_n, \ldots\}$,

- A set of n-ary function symbols ($n \geq 0$), *Func*. 0-ary function symbols are called constants. *Func* contains:

- $\mathcal{A}, \mathcal{B}, \mathcal{Z}, \mathcal{P}, \mathcal{T}, \mathcal{R}$
- a domain-specific set of function symbols, F_{dom}.

In order to reflect the state of other agents with respect to the executed program into the belief base, the following predicates are introduced:

- $\text{In}(a, p, \tau, z)$, defined to hold if and only if $(p, \tau, z) \in \text{PBase}(a)$. This allows an agent to reason about its beliefs about the internal states of other agents. For instance, $\text{Bel}_a \text{In}(b, p, \tau, z)$ denotes that a believes b to be committed to τ in p and that b is currently in state z. Due to the plan base axioms, Σ_p, the occurrence of $\text{In}(a, p, \tau, z)$ within a belief base is limited in the same way as the corresponding plan-task-state triples in a plan base.

- $\text{HasRole}(a, r)$, expressing that agent a holds role r,

- $\text{Succeeded}(a, p, \tau)$, true if and only if agent a successfully completed task τ in plan p.

The set of predicates in $\mathcal{L}(Pred, Func)$, $Pred$ is assumed to contain these predicates. Furthermore, since the rule-based execution of an ALICA program needs additional stateful information, a set of additional predicates P_{rules} is assumed to be a subset of $Pred$. The operational semantics discussed in Section 5.13 assume the following predicates in P_{rules}:

- $\text{Handle}_f(b, z)$, which holds if an agent should handle the failure of behaviour b in state z,

- $\text{Handle}_f(p)$, which is true if an agent should handle the failure of plan p,

- $\text{Failed}(p, i)$, indicating that plan p failed i-times,

- $\text{Failed}(b, z, i)$, indicating that behaviour b failed i-times in state z,

- $\text{Alloc}(z)$, true if and only if an allocation of tasks to agents for state z is deemed necessary,

- $\text{Success}(b)$ and $\text{Fail}(b)$, indicating a success or failure signal from behaviour b, as already noted in Section 4.1.

Finally, P_{dom} contains predicates relating to the world representation of the agent, e.g., $DistanceTo(object, dist)$ or $Carries(agent, object)$. The language elements introduced here allow agents to reason about the environment (with symbols in F_{dom} and P_{dom}), and about the internal states of themselves and each other. Thus, calculations such as role assignment can be done with respect to an agent's beliefs.

In order to capture the relationship between the different predicates reflecting beliefs about internal states of agents, we define a set of axioms, Σ_b.

Definition 5.6. Let Σ_b contain for each agent a in $\mathcal{A}$ the following axioms:

- Unique Name Axioms over agents, behaviours, plans, states, tasks, and roles:

  $UNA(\mathcal{A}, \mathcal{B}, \mathcal{P}, \mathcal{Z}, \mathcal{T}, \mathcal{R})$

- If failure handling for a plan is needed, it is relevant:

  $$\left(\text{Bel}_a \, \text{Handle}_f(p) \vee \text{Failed}(p, i) \right) \rightarrow \left(p = p_0 \vee (\exists p', \tau, z) \, \text{In}(a, p', \tau, z) \right.$$
  $$\left. \wedge p \in \text{Plans}(z) \right)$$

- In the same way, if failure handling for a behaviour is needed, it is relevant:

  $$\left(\text{Bel}_a \, \text{Handle}_f(b, z) \vee \text{Failed}(b, z, i) \right) \rightarrow (\exists p, \tau) \, \text{In}(a, p, \tau, z)$$

- An agent's success in a task is only relevant as long as there is another agent still within the state that contains the corresponding plan:

  $$\text{Succeeded}(a, p, \tau) \rightarrow (\exists z) p \in \text{Plans}(z) \wedge (\exists a', \tau', p') \, \text{In}(a', p', \tau', z)$$

- Task allocation is only needed for a state inhabited by the agent:

  $$\left(\text{Bel}_a \, \text{Alloc}(z) \right) \rightarrow (\exists p, \tau) \, \text{In}(a, p, \tau, z)$$

Definition 5.7 (Common Knowledge). Let Σ_B be the set given by:

$$\Sigma_B = \Sigma_{syn} \cup \Sigma_{dom} \cup \Sigma_b \cup \Sigma_p$$

where Σ_{dom} is a set of domain-specific axioms, describing the domain, Σ_b is the set of belief base axioms according to Definition 5.6, Σ_{syn} is the set of syntactic constraints (see Section 4.5), and Σ_p is the set of plan base axioms according to Definition 5.4. Σ_B is assumed to be common knowledge in $\mathcal{A}$, i.e., $\Sigma_B \wedge \text{MBel}_{\mathcal{A}} \Sigma_B$ holds.

Definition 5.8 (Belief Base). A set of formulae $B \subset \mathcal{L}(Pred, Func)$ is a *belief base* for agent a if and only if

$$\Sigma_B \cup B \not\models \perp$$

and

$$\text{In}(a, p, \tau, z) \in B \leftrightarrow (p, \tau, z) \in \text{PBase}(a)$$

Thus, an agent's belief base reflects its beliefs about the world as well as its beliefs about all other agents' internal states, i.e., plan bases. The above definition results in a belief base that reflects the intuition that an agent always believes it does what it intentionally is doing. Moreover, the belief base is always consistent with respect to Σ_B. By $\text{BelBase}(a)$, we denote the belief base of agent a.

Definition 5.9 (Agent Proof). Let a be an agent, $\mathcal{F}$ be a set of formulae in $\mathcal{L}$, η a substitution, and ϕ a formula. We denote that a proves ϕ with respect to $\mathcal{F}$ and Σ_B by

$$\mathcal{F} \vdash^a_{\mathcal{L},\eta} \phi$$

where η is the computed answer of the proof. If it is clear from the context, we omit a, $\mathcal{L}$, or η.

The proof operator $\vdash$ refers to a theorem proving algorithm in the logic $\mathcal{L}$ and can be exchanged with respect to the domain.

5.2.3 Belief Update

The belief base of an agent is updated frequently, either to accommodate for new sensory data, communication acts, or internal updates. Here, we only treat the last case explicitly. We write $B + F$ to denote that the belief base B is updated by the (finite) set of formulae $F = \{f_1, f_2, \ldots, f_n\}$.

$$B + F \stackrel{def}{=} B + \bigwedge_{f \in F} f$$

$$B - F \stackrel{def}{=} B + \bigwedge_{f \in F} \neg f$$

We require the belief update operator $+$ to satisfy the KM-postulates [83] U1 - U4, and U8, adopted to accommodate for the static common knowledge, Σ_B, and thus $+$ forms an inertial basic update operator (after Lang [92]):

$$\Sigma_B \cup (B + f) \models f \tag{U1}$$

$$\Sigma_B \cup B \models f \rightarrow (B + f) \leftrightarrow B \tag{U2}$$

If $\Sigma_B \cup B$ and f are both satisfiable then $\Sigma_B \cup (B + f)$ is also satisfiable (U3)

If $B \leftrightarrow C$ and $f \leftrightarrow g$ then $B + f \leftrightarrow C + g$ (U4)

$$(B \cup C) + f \leftrightarrow (B + f) \cup (C + f) \tag{U8}$$

Most importantly, the belief base must be consistent at all times. Note that consistency with respect to Σ_b can be easily established by removing literals of the form Handle$_f(p,z)$, Handle$_f(b)$, Failed(p,i), Succeeded(a,p,τ), Alloc(z), Success(b,z), and Fail(b,z).

Assume, for example, an agent a executes a plan p, hence In(a,p,τ,z) holds in BelBase(a). Now, the agent aborts the execution of p, due to a reaction on a higher level in the plan hierarchy. Then BelBase(a) is updated: BelBase$(a)' =$ BelBase(a) − In(a,p,τ,z). If, for example, an allocation for state z was pending, Alloc$(z) \in$ BelBase(a), it is no longer relevant, and therefore Alloc(z) is removed as well.

5.2.4 Execution Set

The execution set E of an agent holds tuples of the form (b,z), where b is a behaviour and z a state, such that b occurs in z ($b \in$ Behaviours(z)). We call state z the context of b. E can be passed to a software component dedicated to the lower-level execution of behaviours. The contained context allows behaviours to query the belief base for agents inhabiting corresponding plans, states, or tasks. The operational semantics must guarantee that if (b,z) occurs in the execution set of a, a plan p and task τ exist such that (p,τ,z) is an element of its plan base, and thus In(a,p,τ,z) is believed by a. We will return to this property in Section 5.14.

5.2.5 Role Set

The role set R $\subseteq \mathcal{R}$ of an agent consists of all roles the agent currently fulfils. Thus, it holds that Bel$_a$ HasRole$(a,r) \leftrightarrow r \in$ R. Roles are assigned by a role allocation algorithm and dictate an agent's preferences for the available tasks. This two-tiered approach simplifies the task allocation problem [21]. Furthermore, decoupling agents and plans using roles and tasks allows for plans to be developed independently of the team composition, which is usually not completely known during development. In general, a globally optimal solution to role allocation is NEXP-time complete [104]. Section 5.6 presents a practical solution to a simplified problem class.

5.3 Locality

Allowing arbitrary conditions to occur in plans would violate the locality principle defined in Section 5.1. We therefore restrict occurrences of the predicate In(a,ρ,τ,z), such that an agent can only refer to its own local view.

Definition 5.10. A formula ϕ is *plan-local* for plan p if and only if for all occurrences of the predicate $\text{In}(a, \rho, \tau, z)$ in ϕ, ρ is the constant p.

Definition 5.11. Locality Requirement: All formulae occurring as precondition or runtime condition of a plan are plan-local to the respective plan.

This definition is somewhat stricter than needed to enforce the locality principle. Completely forbidding references to agents in other plans, as done here, guarantees that each plan can be easily reused in other contexts, a property very useful for planning algorithms as well as human system designers working with highly complex ALICA programs. Furthermore, as we will see in Section 5.12, it allows the calculation of a task allocation of a plan independent of its siblings and children, hence greatly simplifying the task allocation problem.

Note that references to the predicates $\text{Succeeded}(a, p, \tau)$ and $\text{Failed}(p, i)$ are not restricted, and can be used to refer to other plans. Thus, a plan can react to successful or unsuccessful termination of a sub-plan in a situation specific way. Furthermore, the conditions of transitions are not restricted either. Thereby, a transition can freely refer to sub-plans, giving rise to some modelling patterns, which we will discuss in Section 10.1. Since transitions are irrelevant in solving task allocation problems, this does not cause any problems. Conditions of behaviours are treated separately (see Section 5.7).

5.4 Team Configuration

Based on the notion of an agent executing a plan, which is represented in the belief base by the predicate $\text{In}(a, p, \tau, z)$, we introduce the concept of a team working on a plan. In ALICA, each agent constantly monitors the actions of its team members with respect to the plans it participates in. Hence, an agent can not only react to another agent breaking down and consequently being removed from the team, but also each agent considers the progress of all other agents when making a decision, such as committing to a task within a plan. Note that due to locality, an agent might not be able to deduce the state of the team with respect to plans it does not participate in.

Definition 5.12. Let a be an agent in A and let $\text{TeamIn}(A, p)$ denote that team $A \subseteq \mathcal{A}$ executes a plan p, formally:

$$
\begin{aligned}
\text{TeamIn}(A, p) \overset{def}{=} \ & (\forall \tau \in \text{Tasks}(p))(\exists n_1, n_2)\, \xi(p, \tau) = (n_1, n_2) \wedge \\
& (\exists A')A' \subseteq A \wedge n_1 \leq |A'| \leq n_2 \wedge \\
& (\forall a' \in A)a' \in A' \leftrightarrow (\exists z)\, \text{In}(a', p, \tau, z) \vee \text{Succeeded}(a', p, \tau)
\end{aligned}
$$

If $A = \mathcal{A}$, we abbreviate TeamIn$(\mathcal{A}, p)$ by TeamIn(p).

Definition 5.12 captures the most important notion of a team working on a plan. It is non-monotonic in the sense that there can be two sets A, A' with $A' \supset A$ such that TeamIn(A, p) and $\neg$TeamIn(A', p). Hence, an agent's beliefs regarding its team are vital for its evaluation of plans. Note that this definition also takes successful task completions into account.

5.5 Success Semantics

Depending on the domain, some behaviours and tasks can be completed, that is, there is a distinct goal description that is meant to be reached. Should an agent reach that goal description, the behaviour or the task is said to be completed, and the agents in question are free to work on something else. In case of a behaviour, this success is captured by the predicate Success(b), which indicates that the postcondition of the behaviour is reached. An agent a completing a task τ in plan p supports the believe in Succeeded(a, p, τ). As already apparent from Definition 5.12, having already succeeded in a task influences how agents regard the team's state with respect to the corresponding plan.

In various scenarios, one is also interested in the success of a whole plan instead of a single task. However, the success of an agent in a task is not enough to conclude that the corresponding plan is successful. Instead, the success of a plan depends on the state of its tasks. Since some of the tasks of the plan may be optional, a new notion is needed to capture the relationship between a plan's success and its tasks' success.

Consider, for instance, a set of robots tasked with transporting some equipment through unknown terrain. It could be beneficial to have some robots scout ahead for the best route to avoid carrying the equipment over unnecessarily long routes. However, if only a few robots are available, scouting can be omitted in favour of more transporting robots. Hence the plan can still succeed even though no robot ever completed the scouting task.

Definition 5.13. A task can either be optional or required with respect to a plan. We denote the set of tasks which are required for plan p as Required(p). A plan is successful once, for all its required tasks, the minimal number of agents have successfully completed it.

$$\text{Succeeded}(p) \stackrel{def}{=} (\forall \tau \in \text{Required}(p))(\exists n, n', m)\, \xi(p, \tau) = (n, m)$$
$$\wedge n' = \max(1, n) \wedge n' \leq |\{a \mid \text{Succeeded}(a, p, \tau)\}|$$

Hence, for any required task, the minimal number of agents able to execute it must also complete it. In case the minimal cardinality is zero, at least one agent must complete it.

As an example, consider again the transportation scenario above. There could be debris blocking the path of the transporting agents. By using a minimal cardinality of zero, it would be allowed for all agents to put down the cargo, move the debris out of the way in another task and continue transporting without aborting the execution of the plan.

If the cardinality of plan-task pair (p, τ) is for instance $(2,4)$, this entails that at least two agents must complete τ for p to be completed. The actual number of executing agents can of course be higher. Although initially, at least two agents must start working on τ to satisfy TeamIn(p), these are not necessarily those which complete the task.

When agents are working on completing a plan, they do not necessarily complete all required tasks at once. Instead, some tasks may be completed early, freeing the respective agents to do something else. The function $\xi_t : \mathcal{P} \times \mathcal{T} \times 2^{\mathcal{L}_S} \mapsto \mathbb{N}_0 \times (\mathbb{N}_0 \cup \{\infty\})$ captures the currently required plan cardinalities, given a set of beliefs:

$$\xi_t(p, \tau, B) = \begin{cases} (n, m) & \text{if } \tau \notin \text{Required}(p) \\ (\max(0, n - c), \max(0, m - c)) & \text{otherwise} \end{cases}$$

where $\xi(p, \tau) = (n, m)$ and $c = |\{a \mid \text{Succeeded}(a, p, \tau) \in B\}|$. The notion of still required agents, captured by ξ_t is used during task allocation as discussed in Section 5.8.

By the belief axioms Σ_b, the success of a task is only valid until all agents have left the corresponding plan. If the agents leave a plan entirely and reenter it at some later time, the plan must be completed again, i.e., the second execution refers to a new problem instance. Otherwise, once a plan is completed it can never be executed again until the agents are reset by some other means.

5.6 Role Allocation

In Section 4.4, we introduced the basic concept of roles. We now discuss a way of allocating roles to agents. Campbell and Wu [20] give a detailed overview of existing role allocation methods. We do not extend the state-of-the-art in role allocation, but instead demonstrate how existing approaches can be integrated into ALICA by example.

Similar to Vail and Veloso [170], we follow a broadcast-and-compute approach, where agents broadcast information relevant to the role allocation and compute the allocation locally, allowing for a highly reactive allocation. Compared to their work however, roles in ALICA are relatively static as they do not depend on highly dynamic properties of the domain such as positions of robots or other objects of interest. Instead, they solely depend on the number and capabilities of the agents involved. More dynamic features are captured by task allocation, which we will discuss in Section 5.8.

In order to allocate agents to roles, a measurement is needed, a *role utility*, expressing the adequateness of an agent for a certain role. This utility is determined by the required and provided capabilities of the role and the agent, respectively. We assume a similarity measurement $\Delta\colon C \times C \mapsto [0,1]$, such that $\Delta(x,x) = 1$ and $\Delta(x,x^c) = 0$, where x^c denotes the complement of x. Such measurements based on fuzzy sets can be found in [186].

Definition 5.14 (Role Utility). The utility $\mathcal{U}(a,r)$ of a certain agent a for a certain role r is based on the similarity Δ between the provided capability of the agent $\mathrm{Cap}(a)$ and the required capability of the role $\mathrm{Cap}(r)$. The normalised sum of these values reflects the utility of an agent for a certain role:

$$
\mathcal{U}(a,r) = \begin{cases} 0 & \text{if } (\exists c \in \mathrm{Cap}(r)) \\ & \quad \max_{c' \in \mathrm{Cap}(a)} \Delta(c,c') = 0 \\ \frac{1}{|\mathrm{Cap}(r)|} \sum_{c \in \mathrm{Cap}(r)} \max_{c' \in \mathrm{Cap}(a)} \Delta(c,c') & \text{if } \mathrm{Cap}(r) \neq \emptyset \\ 1 & \text{otherwise} \end{cases}
$$

That is, every required capability is matched with the best matching provided capability of the agent.

Given role utilities, a role allocation procedure can be defined, which distributes roles among all agents of the team. We assume this procedure is encapsulated by a *formation*.

Definition 5.15. For a given ALICA program we denote the formation by F. The formation holds the specific role allocation algorithm as well as any further information needed to compute a role allocation. A role allocation is a set of beliefs of the form $\mathrm{HasRole}(a,r)$, computed by $\mathrm{F}(A)$, where A is a set of agents.

A simple yet effective formation can be described by a priority list. A priority list defines a total order over role instances out of $\mathcal{R}$ and assigns the best fitting agent to the most important role instance according to the total order. A corresponding algorithm is depicted in Listing 5.1. In this specific case, if there are

```
A := set of available agents;
R := non−empty sorted list of roles according to their priority ;
i := 0;
j := 0;

while(A ≠ ∅) {
   a := argmax_{a∈A} U(a, r[i]);
   if (U(a, r[i]) > 0) {
     assign a to role r[i];
     remove a from A;
     j := 0;
   } else j++;
   i := (i+1) mod |R|;
   if (j = |R|) return FAILURE;
}
```

Listing 5.1: Priority-Based Role Allocation

more agents than roles, the algorithm starts over with the remaining agents. Also note that the value 0 is used to denote the inability of an agent to take on a certain role, expressing that it lacks a certain capability.

In the robotic soccer domain, a suitable list of role instances could be:

$$\vec{R} = (\textit{Attacker, Goalie, Defender, Attacker, Defender, Supporter})$$

which describes that first and foremost, an attacking robot is needed, followed by a Goalie and a Defender. If further robots are available, a second Attacker and a second Defender are allocated, followed by a Supporter. Capabilities can now be used to express that only suitably large robots take on the goalie role, and only robots able to kick the ball take on the role of an attacker. If a robot cannot be allocated to a role due to missing capabilities, the algorithm fails after allocating all robots which can take on a role.

More sophisticated techniques can handle cardinalities for each role, as well as take interrelated utilities into account. However, we do not discuss these here, as the underlying problem is similar to the task allocation problem discussed in Section 5.8. Note that the role allocation only depends on the number and capabilities of the available agents and thus only needs to be calculated in the event one of these changes, e.g., when an agent loses a capability or a new agent joins the team.

This approach enables the integration of monitoring facilities which detect malfunctioning components of agents or robots. In this case, role allocation can react on the newly missing or degraded capability and allocate the damaged robot to a less critical role. A similar approach was discussed by Weber and Wotawa [177].

In Section 5.13, we provide an operational rule, which integrates role allocation into the runtime semantics of ALICA.

5.7 Canonical Behaviour Plans

Behaviours are the central primitives out of which ALICA programs are constructed. Coordination and cooperation depend on accurate information about which agent is executing which behaviour, whether this information is represented explicitly or implicitly. In ALICA, behaviours encapsulate atomic action programs, treated as black boxes, which are annotated by pre-, post-, and runtime conditions. In contrast to plans, we did not introduce a belief which reflects whether or not an agent executes a behaviour, thus an agent cannot reason about the behaviours executed by another agent directly. However, agents can reason about plans, the correspondingly higher level structure. This gap becomes apparent, when behaviours are annotated with non-trivial pre- and runtime conditions which only a local agent can evaluate.

The gap can be closed by introducing appropriate beliefs which represent the execution of behaviours. However, such beliefs would unnecessarily complicate the operational semantics, leading to poor readability. Instead, we note that behaviours can be embedded in plans, which reflect their conditions. Thereby, a behaviour is lifted such that the team can reason and communicate about the relationship between individual agents and specific behaviours.

This embedding must guarantee that whenever the plan is executed, the behaviour is executed as well. Such an embedding can be achieved automatically by generating a plan and a plantype for each available behaviour. A straightforward embedding is given by a plan with two states: one containing the behaviour, and one terminal state. We refer to this straightforward embedding as a *canonical* plan for a behaviour.

Definition 5.16 (Canonical Plan). A canonical plan for a behaviour b, p_b is defined by the following structure:

- States$(p_b) = \{z_b, z_b^s\}$

- Tasks$(p_b) = \{\tau_0\}$

- Init$(p_b, \tau_0) = z_b$

- $\xi(p_b, \tau_0) = (0, \infty)$

- Pre$(p_b) = $ Pre(b), Run$(p_b) = $ Run$(b) \wedge \neg$Fail(b)

- $\mathcal{U}(p_b)(B) = 1$

- $\text{PlanTypes}(z_b) = \emptyset$

- $\text{Behaviours}(z_b) = \{b\}$

- $\mathcal{W} \ni (z_b, z_b^s, \text{Succeeded}(b))$

- $\text{Success}(p_b) = \{z_b^s\}$, $\text{Fail}(p_b) = \emptyset$

- $\text{Post}(z_b) = \text{Post}(b)$

- $\text{Required}(p_b) = \{\tau_0\}$

The canonical plan is the single element of a corresponding canonical plantype:

Definition 5.17 (Canonical Plantype). A canonical plantype for a behaviour b, P_b contains exactly the canonical plan for b, $P_b = \{p_b\}$

Assuming that each behaviour is embedded in a canonical plan, agents are freed from considering behaviour conditions separately; it suffices to consider plan conditions. While it is possible to use other embeddings, which implement additional structure, this does not yield additional expressiveness. Therefore, we assume in the following that every behaviour only occurs in its canonical plan.

5.8 Task Allocation

One of the most central problems to multi-agent coordination is task allocation [20, 104]. In ALICA, this problem is defined by the choice of a plan from a plantype and an assignment of agents to tasks within the selected plan. As ALICA allows multiple agents per task, the problem is NP-hard [55]. In contrast to task allocation discussed by Nair et al. [104], the utility function of each plan involved potentially depends on the complete allocation, thus the utility for each agent may not be computable without knowing the complete allocation.

The task allocation problem arises whenever an agent enters a state. In this case, it has to decide for all plantypes within that state, which plan to execute and to which, if any, task it commits to. That is, it has to calculate a task allocation for all participating agents. Following a broadcast-and-compute approach to guarantee swift reactions, the agent computes the task allocation by itself based on its current beliefs. Ideally, every agent involved computes the same task allocation, however this is not guaranteed, since the involved agents may have different sensor data affecting their calculations. ALICA follows a weak commitment idea. Calculated allocations are assumed to be correct until contradictory information is available. Chapter 6 will elaborate on how allocations can be repaired and conflicts resolved.

Formally, a task allocation C for a plan p is a set of predicates of the form $\text{In}(a_i, p, \tau, z)$, one for each agent a_i allocated to participate actively in p. For each $\text{In}(a_i, p, \tau, z)$ in C, τ is a task relevant for p, i.e., an element of $\text{Tasks}(p)$, and z is the initial state of τ in p, $\text{Init}(p, \tau)$. An allocation is never calculated without a frame of reference. Typically, a situation in the problem domain, for instance the current situation or a planned hypothetical situation, acts as frame of reference. We represent this frame of reference as a set of sentences of $\mathcal{L}$. In line with the belief update notion, this set of assumptions is also sometimes referred to as a belief state. Note that these assumptions may contain literals of the form $\text{In}(a, p, \tau, z)$ as well, thus some agents might already be allocated to some tasks in p.

Definition 5.18. A *task allocation* is a set of ground beliefs of the form $\text{In}(a, p, \tau, z)$.

Definition 5.19. A task allocation C is *valid* for plan p under the set of assumptions $\mathcal{F}$ if and only if:

- $(\forall l \in C)(\exists a, \tau) l = \text{In}(a, p, \tau, \text{Init}(p, \tau)) \wedge \tau \in \text{Tasks}(p)$

- $\Sigma_B \cup \mathcal{F} \cup C \not\models \bot$

- $\Sigma_B \cup \mathcal{F} \cup C \vdash \text{Pre}(p) \wedge \text{Run}(p)$

- $\Sigma_B \cup \mathcal{F} \cup C \vdash \text{TeamIn}(p)$

- $\mathcal{U}(p)(\mathcal{F} \cup C) \geq 0$

The macro $\text{TAlloc}(p|\mathcal{F})(C)$ is defined to hold if only if C is a valid task allocation for plan p under the assumptions $\mathcal{F}$.

By this definition, a valid task allocation is consistent with the ALICA axioms as well as with the corresponding situation $\mathcal{F}$. If an agent allocates with respect to the current state, $\mathcal{F}$ equals its belief base. Hence, an agent believed to be already committed to a task in p will be considered. This allows for a dynamic repair in case some agents withdraw from a plan and need to be replaced. Furthermore, the definition guarantees that the plan allocated to is properly executed by the team. Finally, since a positive utility is required, the utility function can forbid the execution of certain assignments or execution under specific conditions.

The utility function of a plan indicates not only how well the plan fits a certain situation, but it also takes the task allocation into account. Thus, it can reflect rewards for certain tasks similar to the approach discussed by Nair et al. [104]. An agent allocating tasks for a plan p should maximise $\mathcal{U}(p)(\mathcal{F} \cup C)$, where C is a valid task allocation and $\mathcal{F}$ the assumptions, e.g, the currently believed situation. In other words, it calculates

$$T = \underset{C}{\mathrm{argmax}}\, \mathcal{U}(p)(\mathcal{F} \cup C) \text{ subject to } \mathrm{TAlloc}(p|\mathcal{F})(C)$$

Definition 5.20. A task allocation C is *valid* for a plantype P under the set of assumptions $\mathcal{F}$ if and only if there exists a plan p in P such that $\mathrm{TAlloc}(p|\mathcal{F})(C)$.

By Definition 5.20, a valid task allocation for a plantype is a valid task allocation for one of the plans within that plantype, thus it selects a plan. The optimisation task is extended to involve the selection of plan p as well:

$$T = \underset{C}{\mathrm{argmax}}\, \underset{p \in P}{\max}\, \mathcal{U}(p)(\mathcal{F} \cup C) \text{ subject to } \mathrm{TAlloc}(p|\mathcal{F})(C)$$

Since only valid allocations are considered for this optimisation task, a solution might not exist in all cases.

5.9 Recursive Task Allocation

Based on the notion of task allocation, we discuss a recursive version, which allocates agents to a branch of plans in the plan-tree. Intuitively, whenever an agent enters a state, it must solve the task allocation problem for all plantypes within that state, enter the states entailed by the allocation, and is then confronted with task allocation problems in these sub-states as well.

Due to the hierarchical nature of ALICA programs, the recursive task allocation problem is central to ALICA. In Section 5.13, we will discuss how allocations are adapted dynamically during the execution to accommodate dynamically changing environments or unforeseen problems, such as agents breaking down. Such a dynamic reallocation requires a highly efficient allocation algorithm able to produce optimal allocations in real-time.

The following example illustrates the recursive task allocation problem and we will use it to discuss properties of allocation algorithms.

Example 5.1. *The manager of a larger restaurant wants to use robots to serve his customers. To start out, he buys three serving robots, cleverly named a, b, and c. He specifies a top-level plan for them, P0, and leaves the rest to his staff to complete. Figure 5.1 shows P0 and the plans written by his staff members. For*

simplicity, each plantype involved contains only one plan. Thus, when executing
ServeGuests, a robot can either fetch food from the kitchen and bring it to a ta-
ble (task DeliverOrder*), or observe the customers and take orders if needed (task*
TakeOrder*). The chef of the restaurant wants food to be delivered as fast as possi-*
ble to avoid serving cold food. Thus, he declares that as many robots as possible
should deliver food. Expressing that as a utility function, he writes

$$\mathcal{U}(ServeGuests) = \frac{|\{a \mid \text{In}(a, ServeGuests, DeliverOrder, z)\}|}{100}$$
$$+ 0.1 \frac{|\{a \mid \text{In}(a, ServeGuests, TakeOrder, z)\}|}{100}$$

as he considers 100 *to be the absolute maximum of robots ever serving in his*
restaurant. His smart apprentice specifies a plan for delivering orders to tables,
DeliverOrder*. Knowing that robots will just idly stick around and take up space*
when they are waiting for something to do, he adds a precondition, expressing that
no more robots may execute FetchOrder *than there are dishes ready in the kitchen:*

$$(\exists x) |\text{In}(a, DeliverOrder, Waiter, Z_3)| \leq x \wedge DishesAvail(x)$$

Now consider a situation where the robots just start to work and two dishes are
ready to be served, DishesAvail(2)*. A recursive task allocation for state* Z_0 *in plan*
P0 is due. Maximising the utility of plan ServeGuests*, the only plan within plan-*
type ServeGuestsPT*, would allocate all three robots to the task* DeliverOrder*, thus*
requiring the whole team to execute DeliverOrder*. However, as only two dishes*
are ready to be served, allocating three robots in DeliverOrder *would violate its*
precondition. Thus, only two robots can be allocated to the DeliverOder *task in*
ServeGuests*, the remaining robot should observe the customers and wait for fur-*
ther orders to take.

In this simple example, three robots are allocated to two tasks within the plan
ServeGuests, yielding eight possibilities. In general, the number of possible allocations in a plan with n robots and m tasks is m^n, making a brute force enumeration
of all possibilities quickly infeasible. A more efficient approach is needed. In Section 5.11, we discuss how utility functions can be structured to allow for search
algorithms such as A^* to be employed.

However, allocating robots within a single plan is only part of the problem. The
allocations of the parent plan, *ServeGuests*, and the two sub-plans can interact as
in this example. An algorithm that finds a solution consistent with all conditions
and optimal with respect to the utilities involved would need to consider the global
problem, i.e., consider all plans involved. This would negate any positive effect

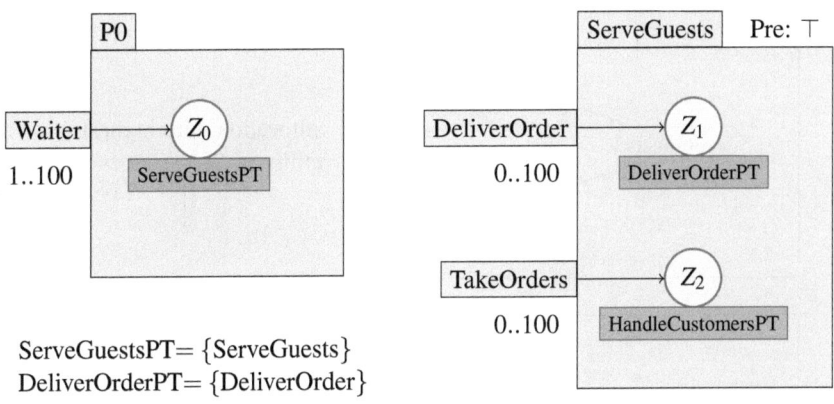

ServeGuestsPT= {ServeGuests}
DeliverOrderPT= {DeliverOrder}

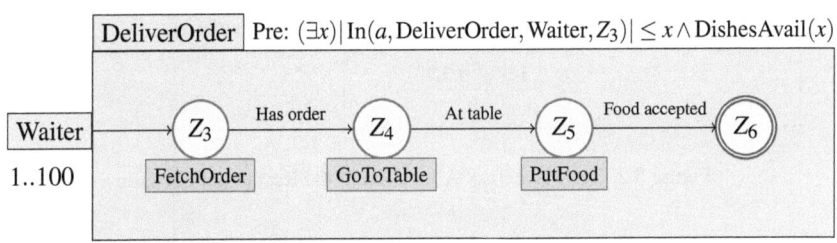

Figure 5.1: Task Allocation Example

the hierarchical decomposition might have on the computation time. Moreover, since each agent needs to calculate the allocation, as communication may be too slow and too unreliable, complex plan structures with many agents can quickly overwhelm the computational power of an individual agent. Finally, the necessary information to decide on an allocation in plan p might not be available to agents not participating in it. Thus, we consider a global allocation algorithm to be infeasible.

In Section 5.1, we introduced the locality principle, which demands that each agent should only be concerned with plans it is involved in and should not do any calculations about plans it does not participate in. Following this principle, the three agents could individually conclude that they cannot allocate all three of them to the task *DeliverOrder*. Using a total order over agents and tasks, each one can individually come to the conclusion that agents b and c should fetch dishes and a should take orders. Agents b and c would evaluate the precondition of *DeliverOrder* to true, while agent a does not evaluate it. Figure 5.2 illustrates this

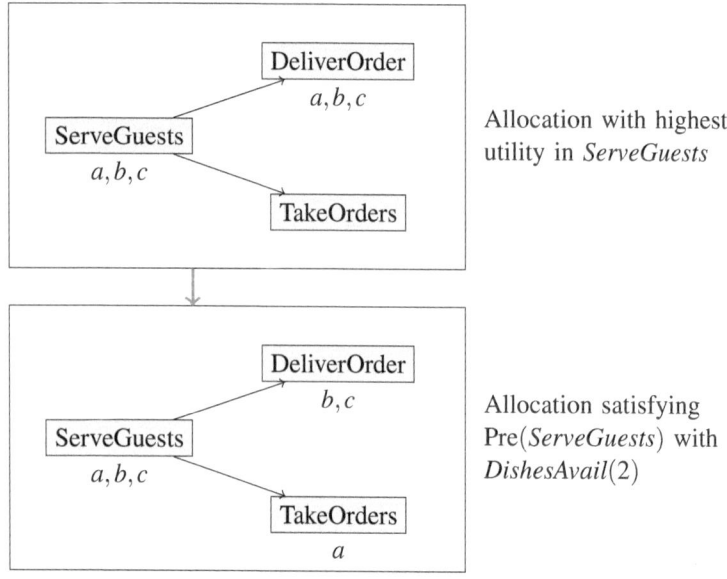

Figure 5.2: Recursive Task Allocation in the Restaurant Scenario

situation. Here, all agents agree on the allocation, and coordination is achieved. However, what if only a single dish is ready to be served? In this case, only one agent is allowed to execute *DeliverOrder*. The ensuing situation is depicted in Figure 5.3.

Given this modified belief, the agents involved calculate different allocations, yielding an incoherent situation. Each robot considers another allocation to be the best valid one with respect to the utility function of *ServeGuests*. At first glance this seems to be an inherent problem of the locality principle, as conditions and utilities of plans at different levels can interact with each other *relative to the domain*, i.e., through arbitrary relations in $\mathcal{L}$.

This effect can be countered in the following ways:

Global Allocation – A global allocation algorithm does not obey the locality principle and thus produces coherent allocations even in situations such as the one above. However, global allocation quickly becomes computationally infeasible as the number of agents, the depth, and the width of the plan-tree increase. Furthermore, as mentioned above, the necessary information to compute a global allocation might not be available to any single agent.

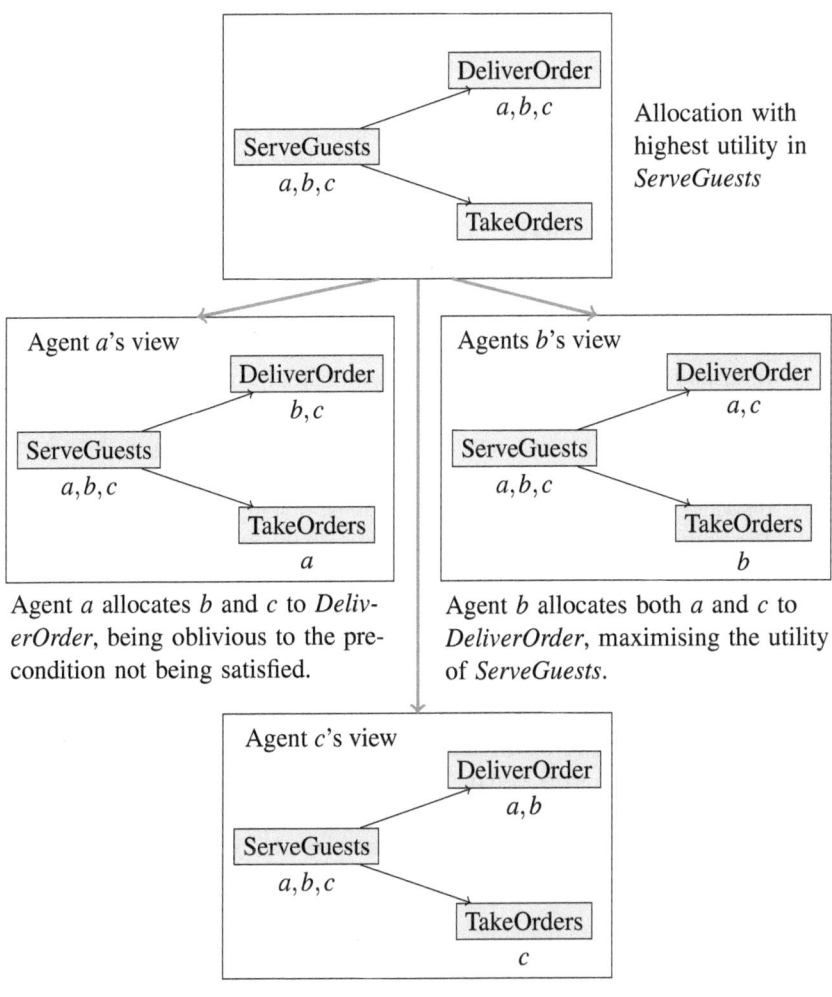

Allocation with highest utility in *ServeGuests*

Agent *a*'s view

Agent *a* allocates *b* and *c* to *DeliverOrder*, being oblivious to the precondition not being satisfied.

Agents *b*'s view

Agent *b* allocates both *a* and *c* to *DeliverOrder*, maximising the utility of *ServeGuests*.

Agent *c*'s view

Maximising the utility, *c* allocates *a* and *b* to *DeliverOrder*.

Figure 5.3: Incoherent Task Allocation in the Restaurant Scenario

Transformation – A computational step before run-time of the plans in question could modify conditions and utilities by incorporating information about sub-plans, thus avoiding situations such as the one above. This process however depends on $\mathcal{L}$, particularly on the decidability of entailment with respect to background knowledge. For first-order logic this is undecidable in general. If the ALICA program is generated by a planning algorithm, this algorithm should take care that the generated programs avoid structures featuring problems such as the above one. In the following, we will discuss what this formally entails.

Communication – Agents can also communicate to arrive at a coherent allocation. In practice, communication takes time and is unreliable, but in some domains might be a reasonable option. An agent could, for instance, inform its team members of its inability to participate in a certain plan and each agent could incorporate this information into the allocation process. In highly dynamic domains, however, it is unclear for how long such an information can be considered valid. In the limit, the information can become outdated before it is received by all agents, making it impossible to arrive at a coherent allocation. Since we are focussing on dynamic domains, we refrain from this option.

Alternatively, agents can elect a leader for each plan, which calculates the local allocation and thus achieves coherence. This is done, for instance, in STEAM [163, 125]. The process of electing a leader consumes some time and introduces an additional problem: when the elected leader breaks down, a new one needs to be elected. Thus, we think that in general the team should be able to work without an explicit leader. In Chapter 6, we discuss how an election process can be triggered upon the detection of a conflict in teamwork and how the resulting leader then takes charge of an allocation problem local to a specific plantype.

Decoupling – The final option avoids the problem of interacting conditions altogether by decoupling the allocation problems. This is achieved by allowing agents to not partake in plans they otherwise would and, instead, idle. Intuitively, this is akin to adding a task to each plan with cardinality $(0, \infty)$ such that conditions and utilities are neutral towards this task. In the restaurant example, this would result in all agents being allocated to the task DeliverOrder, maximising the utility of *ServeGuests*, but the superfluous agents would idle and not partake in the plan *DeliverOrder* until there are dishes available again.

Although the decoupling approach results in allocations which are not optimal with respect to the global problem, it has some appealing properties. Firstly, it guarantees that the agents will agree on an allocation for each plan given their belief bases are equivalent with respect to the conditions and the utilities of each plan *separately*. Secondly, the approach is compatible with both the transformation and the communication options, i.e., incorporating transformations or communication is viable even if the allocation problems are decoupled. Therefore, we will discuss different degrees of coupling between plan and sub-plan allocations and formalise the respective notions under the locality principle in the following.

This aims at the definition of an recursive allocation, which limits the allocation task to those plans, the allocating agent is participating in. In this way, the idea of reducing complexity through hierarchical decomposition is maintained. Recursive task allocation is always done with respect to a state that contains the tree in which agents are to be allocated.

Firstly, we define a minimal requirement an allocation must fulfill, followed by stricter definitions, which enforce a tighter coupling between the allocation of a plan and the allocations of sub-plans.

Definition 5.21 (Recursively valid task allocation). Let C be a task allocation done by agent a for state z under assumptions $\mathcal{F}$, let $\vec{P}$ be the set of plans mentioned in C, i.e., $\vec{P} = \{p | (\exists a', \tau, z') \operatorname{In}(a', p, \tau, z') \in C\}$, and let $\mathcal{G}$ denote $\mathcal{F} \cup C$. Then C is *recursively valid* if and only if:

$$\Sigma_B \cup \mathcal{G} \not\models \bot \tag{5.4}$$

$$\Sigma_B \cup \mathcal{G} \vdash \left(\bigwedge_{p \in \vec{P}} \operatorname{Pre}(p) \wedge \operatorname{Run}(p) \right) \tag{5.5}$$

$$(\forall p \in \vec{P})\mathcal{G} \vdash \operatorname{TeamIn}(p) \tag{5.6}$$

$$(\forall p \in \vec{P})p \in \operatorname{Plans}^+(z) \wedge \left(p \in \operatorname{Plans}(z) \vee (\exists p', \tau', z') \right. \tag{5.7}$$
$$\left. \operatorname{In}(a, p', \tau', z') \in \mathcal{G} \wedge p \in \operatorname{Plans}(z') \right)$$

$$(\forall l \in C)(\exists a', p, \tau)l = \operatorname{In}(a', p, \tau, \operatorname{Init}(p, \tau)) \tag{5.8}$$

$$(\forall p \in \vec{P})\mathcal{U}(p)(\mathcal{G}) \geq 0 \tag{5.9}$$

$$(\forall a')(\exists p', \tau', z') \operatorname{In}(a', p', \tau', z') \in C \rightarrow \mathcal{F} \vdash (\exists p'', \tau'') \operatorname{In}(a', p'', \tau'', z) \tag{5.10}$$

Let $\operatorname{TAlloc}^*(a, z | \mathcal{F})(C)$ denote that C is a recursively valid task allocation for state z done by agent a with respect to the assumptions $\mathcal{F}$.

By Condition 5.4, a recursively valid task allocation is consistent with the assumptions $\mathcal{F}$ and the common knowledge. Hence, by Definition 5.3 and 5.8, an agent cannot be assigned two conflicting tasks. In particular, an agent already believed to be participating in a plan p cannot be reassigned within that plan. Condition 5.5 ensures that all preconditions and runtime conditions of the involved plans are met. Condition 5.6 rules out partially assigned plans, i.e., enforces that each plan is executed by a proper number of agents. Condition 5.7 limits the task allocation to plans mapped onto by plantypes occurring in states the allocating agent enters by adopting the allocation.

Due to Condition 5.8 each agent that is newly allocated to a task is believed to be in the corresponding initial state, i.e., cannot be allocated to arbitrary states within the task. Condition 5.9 ensures that all resulting utility values are not less than 0 and hence the allocation is not considered harmful. Finally, by Condition 5.10, agents can only be allocated if they inhabit the state z, for which the allocation is computed.

The task of finding an optimal solution to this problem is called the *recursive task allocation problem*. We will discuss it in the next section.

Note that the empty set is a recursively valid task allocation. Moreover, a recursively valid task allocation is not required to guarantee execution of all involved plantypes. This allows for plans which are not executable to be skipped and for agents to idle while the rest of the team completes a task if they cannot contribute.

A stricter requirement is enforced by the next property, *completeness*.

Definition 5.22. A recursive task allocation C done by agent a for state z under the assumptions $\mathcal{F}$ is *recursively complete* if and only if it is recursively valid and

$$
\begin{aligned}
(\forall P \in \mathrm{PlanTypes}(z))(\exists p \in P)(\mathcal{F} \cup C \vdash \mathrm{TeamIn}(p)) \wedge \qquad (5.11) \\
(\forall \mathrm{In}(a, p', \tau, z') \in C)(\forall P \in \mathrm{PlanTypes}(z')) \\
\big((\exists p'' \in P)(\mathcal{F} \cup C \vdash \mathrm{TeamIn}(p''))\big)
\end{aligned}
$$

By $\mathrm{TAlloc}_c^*(a, z | \mathcal{F})(C)$ we denote that C is a recursively complete task allocation for state z done by agent a under the assumptions $\mathcal{F}$.

A complete recursive task allocation requires that for the original state z and for every state the local agent a is allocated to, the team executes one plan per contained plantype. In other words, if this property is maintained at all times, the agent must ensure that for each state z it inhabits, for all plantypes $P \in \mathrm{PlanTypes}(z)$, one plan $p \in P$ is executed. Note, that the agent a does not necessarily participate in the execution of all plans, it just has to make sure the team executes them. This is akin to a Joint Intention [97], all involved agents commit to the execution of the plan, even if some of them do not actively participate. In case an active participant breaks down and the execution of the plan is endangered, the passive participants are forced to become active due to their commitment.

While enforcing the completeness property does not cause any problems in the scenario described above, it can cause conflicting team states whenever an agent or a group of agents cannot execute a sub-plan, but is so required by an allocation on an ancestor level.

An even tighter coupling of the individual allocation problems is captured by the notion of a *perfect allocation*:

Definition 5.23. A recursive task allocation C done by agent a for state z under the assumptions $\mathcal{F}$ is *recursively perfect* if and only if it is recursively valid and

$$(\forall P \in \text{PlanTypes}(z))\,\text{In}(a',p',\tau,z) \in \mathcal{F} \cup C \to \qquad (5.12)$$
$$(\exists z',p'',\tau')p'' \in P \wedge \text{In}(a',p'',\tau',z') \in \mathcal{F} \cup C$$
$$\text{In}(a,p,\tau,z') \in C \to ((\forall a')\,\text{In}(a',p,\tau,z') \in C \to \qquad (5.13)$$
$$(\forall P \in \text{PlanTypes}(z'))(\exists p',\tau',z'')p' \in P$$
$$\wedge \text{In}(a',p',\tau',z''))$$

Analogous to the above definitions, let $\text{TAlloc}_p^*(a,z|\mathcal{F})(C)$ denote that C is a recursively perfect task allocation for state z done by agent a under the assumptions $\mathcal{F}$.

The conditions 5.12 and 5.13 require the task allocation to allocate all agents. That is, every agent inhabiting the original state z is part of the execution of every plantype in z (Condition 5.12). Furthermore, by Condition 5.13, the same holds for the recursive case. If a is allocated so that it inhabits a certain state, all agents in that state must participate in the execution of all plantypes within that state.

It is easy to see that due to Condition 5.6 of Definition 5.21, a recursively perfect allocation is also recursively complete. Both the complete and the perfect allocation can suffer from the coupling problem, where utilities and conditions of plans on different levels interact in such a way that the agents cannot calculate a corresponding allocation individually.

In a sense, plans ought to exhibit a certain compatibility in order to avoid such interactions. Such a notion of compatibility depends on the domain axioms, Σ_{dom}.

Definition 5.24. A plantype P is *hierarchically compatible* with task τ in plan p if and only if for all belief bases $\mathcal{F}$, if a valid allocation C for p exists under $\mathcal{F}$, a valid task allocation C' for a plan $p' \in P$ under $\mathcal{F} \cup C$ exists as well, such that $\{a \mid \text{In}(a,p,\tau,\text{Init}(p,\tau)) \in \mathcal{F} \cup C\} \supseteq \{a \mid \text{In}(a,p',\tau',z') \in \mathcal{F} \cup C'\}$.

Definition 5.25. A plan p is *sound* with respect to Σ_{dom}, if and only if

- for all tasks $\tau \in \text{Tasks}(p)$, all plantypes in $\text{PlanTypes}(\text{Init}(p, \tau))$ are hierarchically compatible with τ in p, and

- for all states $z \in \text{States}(p)$, all plans in $\bigcup \text{PlanTypes}(z)$ are sound.

Hence, during recursive allocation, in a compatible plan, a valid allocation can be derived using at most the available agents. The soundness property can guarantee agreement of the team with respect to calculated recursive allocations, given that the belief bases of the agents involved are not contradictory.

The notion of sound plans does not necessarily yield perfect allocations, this requires a slightly stronger notion, captured by the next two definitions.

Definition 5.26. A plantype P is *hierarchically perfectly compatible* with task τ in plan p if and only if for all belief bases $\mathcal{F}$, if a valid allocation C for p exists under $\mathcal{F}$, a valid task allocation C' for a plan $p' \in P$ under $\mathcal{F} \cup C$ exists as well, such that $\{a \mid \text{In}(a, p, \tau, \text{Init}(p, \tau)) \in \mathcal{F} \cup C\} = \{a \mid \text{In}(a, p', \tau', z') \in \mathcal{F} \cup C'\}$.

Definition 5.27. A plan p is *perfectly sound* with respect to Σ_{dom}, if and only if

- for all tasks $\tau \in \text{Tasks}(p)$, all plantypes in $\text{PlanTypes}(\text{Init}(p, \tau))$ are hierarchically perfectly compatible with τ in p, and

- for all states $z \in \text{States}(p)$, all plans in $\bigcup \text{PlanTypes}(z)$ are perfectly sound.

Thus, in a perfectly compatible plan, all available agents can be allocated, yielding a perfect allocation.

The above definitions allow for various options of performing recursive task allocation, from the base case where the result only needs to be valid, to the strongest requirement of perfect allocations. Given that the involved plans are sound or even perfectly sound, corresponding results can be guaranteed, as we will show in Section 5.12.

Whether soundness or perfect soundness can be established depends on the domain, and the generation process. A development process for ALICA programs can include a validation step, checking the modelled plans for the respective properties, while a planning or learning algorithm could be constrained to only produce sound or perfectly sound plans. In general, an unsound plan can be transformed into a sound one by specialising its conditions such that they entail the conditions of the sub-plans. Of course, this requires that all agents involved at the parent level have the necessary information available.

In the following, we discuss the notion of an optimal allocation, which maximises the utility functions involved. In Section 5.12, we will derive a corresponding task allocation algorithm.

5.10 Optimal Task Allocation

In the previous section, we formally defined the task allocation problem as well as valid solutions to it. Here, we concentrate on what an optimal solution is and how it can be calculated.

Nair et al. [104] discussed this problem in detail, although they referred to it as role allocation, and do not separate roles and task in the way we do. In their view, a task allocation considers potential rewards stemming from events or action outcomes. A globally optimal task allocation would consider future reallocations due to agents breaking down or the situation changing. They showed that this problem is NEXP-time complete and thus intractable. A locally optimal task allocation only considers the current situation, and is still NP-hard. They also showed how local task allocation algorithms can be improved using a pre-runtime calculation by considering potential future triggers for reallocation. However, in complex domains, this process quickly becomes infeasible as well, as it scales exponentially with the product of the number of triggers considered and the number of available policies. Furthermore, Nair et al. only considered independent rewards, e.g., they assume that a reward can be associated with an agent taking on a certain task. We allow for interrelated utility functions, that depend on the full allocation for the corresponding plan. Moreover, in highly dynamic domains, necessary probability estimates to consider future rewards are hard or even impossible to obtain. Finally, in order to simplify the problem, we consider the hierarchical case, where on each level different utility functions govern the allocation process. These utility functions do not necessarily relate, but can be derived from each other.

In order to design a suitably general and efficient allocation algorithm we establish the following premises:

- The reward problem is left open. A utility function may consider rewards, but this is domain-specific.

- Allocations can occur very frequently and thus must be highly efficient.

- During runtime, the utility of a plan must be fully computable given a task allocation and corresponding situation. It does not depend on allocations for sub-plans. Such dependencies can be removed in a pre-runtime calculation step.

To this end, we establish *hierarchical optimality* as a criteria for a solution to the recursive task allocation problem.

Definition 5.28. A recursive task allocation C for state z by agent a given the assumptions $\mathcal{F}$ is hierarchically optimal, if and only if C is the smallest set such that for all plantypes P mentioned in C, i.e., $(\exists a', p, \tau, z')p \in P \wedge \text{In}(a', p, \tau, z') \in C$:

$$\{\text{In}(a', p', \tau', z') \mid \text{In}(a', p', \tau', z') \in C \wedge p' \in P\} = \underset{D}{\text{argmax}} \, \underset{p \in P}{\text{max}} \, \mathcal{U}(p)(\mathcal{F} \cup D)$$

subject to either $\text{TAlloc}_c^*(a, z|\mathcal{F})(C)$ or $\text{TAlloc}_p^*(a, z|\mathcal{F})(C)$.

That is, a hierarchically optimal task allocation maximises the utility of every involved plantype while maintaining completeness or perfect completeness, depending on which of the two properties are required. Note that, if only validity is required, i.e., only $\text{TAlloc}^*(a, z|\mathcal{F})(C)$ is enforced, the empty allocation becomes optimal, as their is no incentive for the agents to take on any plan. Without the completeness requirement, such an incentive would need to select plans to execute. However, the selection of plans to execute should be governed by the plan structure itself (potentially supplied by a planning algorithm) and the state of the team within it. Thus, we see such an additional selection mechanism as superfluous and thereby establish completeness as the minimal required property for recursive task allocations.

5.11 Utility Functions

Besides preconditions and runtime conditions, utility functions capture the applicability of a plan in a certain situation, and determine a task allocation. Thus, they fulfil a pivotal role within ALICA. As mentioned in the previous section, calculating a task allocation calls for an efficient search. Thus we impose a structure on utility functions, which allows for a heuristic estimate.

Firstly, a utility function depends on both the current situation, captured by a domain-specific part, and the current role assignment, captured by a domain-independent function.

Definition 5.29. The function $\text{pri}(p)$ evaluates the preferences of all agents involved in executing plan p towards their current tasks in p:

$$\text{pri}(p)(\mathcal{F}) = \begin{cases} -1 & \text{if } \mathcal{F} \vdash (\exists a, \tau, r)\phi[a, p, \tau, r] \wedge \\ & \qquad \text{Pref}(r, \tau) < 0 \\ \frac{1}{|A|} \sum_{\mathcal{F} \vdash \phi[a, p, \tau, r]} \text{Pref}(r, \tau) & \text{otherwise} \end{cases}$$

where $\phi[a, p, \tau, r] = \text{HasRole}(a, r) \wedge (\exists z) \text{In}(a, p, \tau, z)$.

That is, $\mathrm{pri}(p)$ sums up all preferences of allocated agents towards their corresponding task. The factor $\frac{1}{|\mathcal{A}|}$ normalises the result. In case a single preference is below zero, $\mathrm{pri}(p)$ is defined to be -1, following the idea that negative preferences express an inability to do something.

Definition 5.30 (Utility Function). The utility $\mathcal{U}(p)(\mathcal{F})$ of a plan p with respect to belief base $\mathcal{F}$ has the form:

$$\mathcal{U}(p)(\mathcal{F}) = \begin{cases} -1 & \text{if } \mathrm{pri}(\mathcal{F}) < 0 \\ w_0 \, \mathrm{pri}(\mathcal{F}) + \sum_{1 \le i \le n} w_i f_i(\mathcal{F}) & \text{if } \mathcal{F} \models \mathrm{TeamIn}(p) \\ 0 & \text{otherwise} \end{cases}$$

A utility function is a weighted sum of several functions $\mathrm{pri}, f_1, \ldots, f_n$ over belief sets. The weights w_i are constants such that $\sum_{i=0}^{n} w_i = 1$ and $(\forall i) 0 \le w_i \le 1$. The functions $f_i \colon 2^{\mathcal{L}_S} \mapsto [0..1]$ capture domain-specific information. Following the locality principle introduced in Section 5.1, in all occurrences of the predicate $\mathrm{In}(a, \rho, \tau, z)$ in each f_i, ρ may only refer to p.

By this definition, utility functions are restricted to weighted sums. However, as w_0 can be set to 0 and each summand can be arbitrarily complex, this structure is not confining. The next section will show that it is good practice to make use of this structure and keep each summand as simple as possible.

5.12 Task Allocation Algorithm

A task allocation algorithm computes an assignment of agents to tasks, represented as a set of beliefs. The result should satisfy Definition 5.20 or Definition 5.28 in case of a hierarchical allocation.

We propose an algorithm based on the well-known search algorithm A^* by Hart et al. [67]. Though in contrast to the classical definition, we are maximising a utility function instead of minimising a cost function. For this, we derive a heuristic estimate H of the utility function $\mathcal{U}$. In each iteration, the search procedure will expand a search node by assigning an agent to all possible tasks. An allocation is found once all agents are assigned and the plan conditions hold given the calculated allocation. As a utility function maps the belief state of an agent onto the real numbers, a heuristic function maps the belief state together with a set of still unassigned agents to the real numbers: $H(p) \colon 2^{\mathcal{L}_S} \times 2^{\mathcal{A}} \mapsto \mathbb{R}$. The efficiency of A^* is determined by both the accuracy and the computing time of the heuristic.

Given a utility function according to Definition 5.30, we can derive a heuristic in the following way: For a given search node for a search under the assumptions $\mathcal{F}$ with partial allocation H and available agents A, on the path to the goal allocation G, it holds that

$$\mathcal{U}(p)(\mathcal{F} \cup G) \leq \max_{H'} \mathcal{U}(p)(\mathcal{F} \cup H \cup H')$$

where H' allocates exactly all agents in A. Further,

$$\max_{H'} \mathcal{U}(p)(\mathcal{F} \cup H \cup H') \leq w_0 \max_{H'} \mathrm{pri}(\mathcal{F} \cup H \cup H') + \sum_{1 \leq i \leq n} w_i \max_{H'} f_i(\mathcal{F} \cup H \cup H')$$

due to $w_i \geq 0$. That is, by maximising each summand individually, we construct an admissible heuristic. Note that again in contrast to the classical definition, our heuristic includes the utility, thus it estimates the complete utility of a partial node instead of the potential gain. This is merely a technical way to simplify the definitions below.

In case only complete allocations are required, the summand $\max_{H'} \mathrm{pri}(\mathcal{F} \cup H \cup H')$ can be further estimated by

$$h_{\mathrm{pri}}(p)(\mathcal{F}, A) = \begin{cases} -1 & \text{if } \Phi[p, \mathcal{F}, A] \\ \frac{1}{|A|} \left(\sum_{\mathcal{F} \vdash \phi[a,p,\tau,r]} \mathrm{Pref}(r, \tau) + \right. \\ \left. \sum_{a \in A} \max_{\tau: \psi[a,p,\tau,r]} \mathrm{Pref}(r, \tau) \right) & \text{otherwise} \end{cases}$$

where

$$\Phi[p, \mathcal{F}, A] \overset{def}{=} \mathcal{F} \vdash (\exists a \in A, r) \,\mathrm{HasRole}(a, r) \wedge (\exists z, \tau) \,\mathrm{In}(a, p, \tau, z) \wedge \mathrm{Pref}(r, \tau) < 0$$

$$\phi[a, p, \tau, r] \overset{def}{=} \mathrm{HasRole}(a, r) \wedge (\exists z) \,\mathrm{In}(a, p, \tau, z)$$

$$\psi[a, p, \tau, r] \overset{def}{=} \mathcal{F} \vdash \mathrm{HasRole}(a, r) \wedge \tau \in \mathrm{Tasks}(p) \wedge$$
$$(n, m) = \xi_{\mathrm{t}}(p, \tau, \mathcal{F}) \wedge m < |\{a' \mid \mathrm{In}(a', p, \tau, z) \in \mathcal{F}\}|$$

That is, $h_{\mathrm{pri}}(p)(\mathcal{F}, A)$ sums up the maximal preference each unallocated agent has towards any available task in $\mathrm{Tasks}(p)$.

If the resulting allocations ought to be perfectly complete, a stricter heuristic can be used, with

$$\Phi[p, \mathcal{F}, A] \overset{def}{=} \mathcal{F} \vdash (\exists a, r) \,\mathrm{HasRole}(a, r) \wedge (\exists z, \tau) \,\mathrm{In}(a, p, \tau, z) \wedge \mathrm{Pref}(r, \tau) < 0 \vee$$
$$(\exists a \in A) \max_{\tau: \psi[a,p,\tau,r]} \mathrm{Pref}(r, \tau) < 0$$

Such that the heuristic reflects that unallocated agents must be assigned to a task for which their roles have a positive preference.

Trivially,

$$\max_{H'} \text{pri}(\mathcal{F} \cup H \cup H') \leq h_{\text{pri}}(p)(\mathcal{F}, A)$$

$h_{\text{pri}}(p)(\mathcal{F}, A)$ can overestimate the possible value, since it disregards the plan's conditions. Note that the heuristic for the complete allocation problem is also admissible for the perfectly complete allocation problem, but not vice versa.

The domain-specific functions $f_i \colon 2^{\mathcal{L}_S} \mapsto \mathbb{R}$ are estimated by corresponding heuristic functions $h_i \colon 2^{\mathcal{L}_S} \times 2^{\mathcal{A}} \mapsto \mathbb{R}$. As the functions f_i can be arbitrarily chosen, it is difficult to give a constructive definition for their respective heuristic functions.

Assuming T_i is the set of tasks referred to in f_i, then h_i can be defined by:

$$h_i(p)(\mathcal{F}, A) = \max_{Q : \phi[Q, A, \mathcal{F}, p]} f_i(p)(\mathcal{F} \cup Q)$$

where

$$\begin{aligned}
\phi[Q, A, \mathcal{F}, p] = & Q \subseteq \{\text{In}(a, p, \tau, \text{Init}(p, \tau)) \mid a \in A \wedge \tau \in T_i\} \wedge \\
& \big(\text{In}(a, p, \tau, z) \in Q \wedge \text{In}(a, p, \tau', z') \in Q \to \tau = \tau' \wedge z = z'\big) \wedge \\
& (\forall \tau \in T_i)(\exists n, m)\, \xi_t(p, t, \mathcal{F}) = (n, m) \wedge \\
& |\{a \mid (\exists z)\, \text{In}(a, p, \tau, z) \in Q \cup \mathcal{F}\}| \leq m
\end{aligned}$$

Ideally, each function f_i depends on as few tasks as possible, thus allowing simple, yet efficient heuristics. While a heuristic constructed in such a way is admissible, in most cases a more informed heuristic can perform significantly better by exploiting the structure of f_i itself.

Consider the following examples:

Example 5.2. *The summand used by the chef in Example 5.1,*

$$f(\textit{ServeGuests})(\mathcal{F}) = \frac{|\{a \mid \text{In}(a, \textit{ServeGuests}, \textit{DeliverOrder}, z) \in \mathcal{F}\}|}{100}$$

can trivially be extended to the heuristic

$$h(\textit{ServeGuests})(\mathcal{F}, A) = \frac{|\{a \mid \text{In}(a, \textit{ServeGuests}, \textit{DeliverOrder}, z) \in \mathcal{F} \vee a \in A\}|}{100}$$

Yielding a heuristic much more efficient than a construction using the definition above.

In the soccer domain, a common approach is to have the robot closest to the ball take on the task Attack. This can be expressed by the function:

$$f(p)(\mathcal{F}) = \max_{a:\text{In}(a,p,Attack,z)\in\mathcal{F}} 1 - \frac{Dist(a,ball)}{maxDist}$$

where Dist refers to the Euclidean distance between two objects and maxDist to the maximum distance possible, e.g., the diagonal of the soccer field.

The heuristic estimate can be defined by a simple extension to f

$$h(p)(\mathcal{F},A) = \max_{a:\phi[a,p,\mathcal{F},A]} 1 - \frac{Dist(a,ball)}{maxDist}$$

where

$$\phi[a,p,\mathcal{F},A] = \text{In}(a,p,Attack,z) \in \mathcal{F} \vee (\xi_t(p,Attack,\mathcal{F}) = (n,m)$$
$$\wedge m < |\{a' \mid \text{In}(a',Attack,p,z) \in \mathcal{F}\}| \wedge a \in A)$$

In another situation, when a pass can be played, the distance between the passing robot and the receiving robot plays a role, thus a useful function might be:

$$f(p)(\mathcal{F}) = \max_{a:\text{In}(a,p,Passing,z)\in\mathcal{F}} \max_{b:\text{In}(b,p,Receiver,z)\in\mathcal{F}} 1 - \left(\frac{7\,\text{m} - Dist(a,b)}{maxDist}\right)^2$$

specifying the ideal distance for a pass as 7 m.

The corresponding heuristic function is slightly more complex:

$$f(p)(\mathcal{F},A) = \max_{a:\phi[a,p,\mathcal{F},A]} \max_{b:\psi[a,b,p,\mathcal{F},A]} 1 - \left(\frac{7\,\text{m} - Dist(a,b)}{maxDist}\right)^2$$

with

$$\phi[a,p,\mathcal{F},A] = \text{In}(a,p,Passing,z) \in \mathcal{F}$$
$$\vee \xi_t(p,Passing,\mathcal{F}) = (n,m) \wedge m < |\{a' \mid \text{In}(a',Passing,p,z) \in \mathcal{F}\}|$$
$$\wedge a \in A$$
$$\psi[a,b,p,\mathcal{F},A] = \text{In}(b,p,Receiver,z) \in \mathcal{F} \vee (\xi_t(p,Receiver,\mathcal{F}) = (n,m) \wedge$$
$$m < |\{a' \mid \text{In}(a',Receiver,p,z) \in \mathcal{F}\}| \wedge b \in A \wedge b \neq a)$$

These examples illustrate that, while in some cases the automatic construction of heuristic functions described above yields decent results, in other cases, a much more efficient heuristic can easily be constructed by hand. Especially with larger numbers of tasks involved per summand, efficient heuristics become crucial. We will not delve further into the automatic construction of heuristic functions, the interested reader is referred to [120, 123].

Given all individual heuristic estimates, we can formulate the complete heuristic function:

$$H(p)(\mathcal{F},A) = \begin{cases} -1 & \text{if } h_{\text{pri}}(\mathcal{F},A) < 0 \\ w_0\,h_{\text{pri}}(\mathcal{F},A) + \sum_{1 \le i \le n} w_i\,h_i(\mathcal{F},A) & \text{if } \phi[p,\mathcal{F},A] \\ 0 & \text{otherwise} \end{cases}$$

where

$$\phi[p,\mathcal{F},A] = |A| \ge \sum_{n:\tau \in \text{Tasks}(p) \wedge \xi_t(\tau,p,\mathcal{F})=(n,m)} \max(0, n - |\{a' \mid \text{In}(a',p,\tau,z) \in \mathcal{F}\}|)$$

$$\wedge |A| \le \sum_{n:\tau \in \text{Tasks}(p) \wedge \xi_t(\tau,p,\mathcal{F})=(n,m)} m - |\{a' \mid \text{In}(a',p,\tau,z) \in \mathcal{F}\}|$$

Based on utility function and heuristic function, we can define a search based on A^*. The task allocation algorithm for a plantype is depicted in Listing 5.2. We assume a total order over agents, tasks, and plans, which is needed to guarantee that given the same belief state, different agents compute the same allocation, even if the utility of different allocations is the same.

The algorithm consists of two main functions, **InitTaskAlloc** and **NextAlloc**. **InitTaskAlloc** initialises the search queue with one node per plan in the respective plantype. Each node contains a partial allocation and an ordered list of agents free to be allocated. Agents believed to be not already allocated in one of the plans are added to the list of free agents of each node.

NextAlloc iteratively expands the best node according to the heuristic until an allocation is found satisfying **AllocGoal** or the queue is empty, in which case a failure is returned. Subsequent calls to **NextAlloc** will return all valid task allocations for the plantype given the assumptions $\mathcal{F}$ in descending order of their utility.

The function **Expand**, depicted in Listing 5.3, reflects the successor relationship between search nodes and produces the set of children of a search node. Depending on whether or not the overall problem is to find a perfectly complete allocation, the function can be used to enforce all agents to participate or allow for agents to be skipped.

Note that allocations of different plans in the same plantype are in direct competition, even though the utility functions might differ. Hence, the problem of selecting a plan from a plantype is made part of the task allocation problem. This

```
InitTaskAlloc(P,F,Agents) {
  Queue_P := empty Queue;
  foreach (Plan  p  in  P) {
    new Node n;
    n.Plan := p;
    foreach (Agent a  in  Agents) {
        if (F  contains  In(a,p,t,z)  for  some t,z) {
            Remove a from Agents;
        }
    }
    Add n to  Queue_P;
  }
  foreach (Node n in  Queue_P) {
    n.AgentsAvail := Agents;
    n.Alloc := ∅;
    n. Heuristic := H(n.Plan)(F, n.AgentsAvail );
  }
  Sort  Queue_P by heuristic ;
}

NextAlloc(P,F) {
 while (Queue_P is not  empty) {
    n := RemoveFirst(Queue_P);
    if  (AllocGoal(n,P,F)) return  n.Alloc;
    Add Expand(n) to Queue_P;
    Sort  Queue_P by heuristic ;
 }
 return  FAILURE;
}

AllocGoal(node,P,F) {
 if (node.AgentsAvail  not  empty or node. Heuristic  < 0) return false;
 foreach (Task  t  in  node.Plan) {
    (n_min,n_max) := ξ_t(node.Plan,t,F);
    n := |{a  |  In(a,node.Plan,t,z)  ∈ node.Alloc ∪ F}|;
    if  (n < n_min ∨ n > n_max) return false;
 }
 if  (F ∪ node.Alloc ⊢ Pre(node.Plan) ∧ Run(node.Plan)) return true;
 else  return   false ;
}
```

Listing 5.2: Task Allocation Algorithm for a Plantype

requires the utility functions of the plans to be comparable with each other. The structure enforced on utility functions and the fact that they map valid allocations to the interval $[0,1]$ make it easier to design comparable utility functions.

```
Expand(node,F) {
  if (node.AgentsAvail is empty) return ∅;
  a := RemoveFirst(node.AgentsAvail);
  nodes := ∅;
  foreach(Task t in node.Plan) {
    n := copy of node;
    Add In(a,node.Plan,t,Init(node.Plan,t)) to n.Alloc;
    (nmin,nmax) := ξt(node.Plan,t,F);
    n := |{a | In(a,node.Plan,t,z) in n.Alloc ∪ F}|;
    if (n ≤ nmax ∧ n + |n.AgentsAvail| ≥ nmin) {
      n.Heuristic := H(n.Plan)(F ∪ n.Alloc,n.AgentsAvail);
      Add n to nodes;
    }
  }
  if (perfectly complete allocation not required) {
    n := copy of node;
    n.Heuristic := H(n.Plan)(F ∪ n.Alloc,n.AgentsAvail);
    Add n to nodes;
  }
  return nodes;
}
```

Listing 5.3: Node Expansion Function for Task Allocation

Proposition 1. *If the task allocation algorithm returns an allocation for plan p, it is valid.*

Proof. The proof is straight forward. Firstly, if an allocation is returned, it satisfies the conditions defined by **AllocGoal**. Hence, its utility is larger than 0, and thus it satisfies TeamIn(p) for plan p since the cardinalities are satisfied. Further, it satisfies the pre- and runtime conditions of p. As the only beliefs contained in the allocation returned are of the form In$(a, p, \tau, \text{Init}(p, \tau))$ for some agent a, and some task τ in Tasks(p), and agents are only allocated if they are not allocated according to F, the result is also consistent with F and Σ_B.   □

Proposition 2. *If a valid task allocation for a plantype P under the assumptions F using agents A exists, it is eventually returned by **NextAlloc(P,F)**.*

Proof. Proof by contradiction: Assume a valid task allocation C is not returned. C would allocate agents A to a plan $p \in P$. Thus, C is either not expanded or does not satisfy the conditions in **AllocGoal**. Since by the definition of **InitTaskAlloc**, an initial node for plan p is created, and since **Expand** subsequently allocates all available agents to tasks in p, if the cardinalities allow so, C must eventually

```
Let a be the  calculating  agent
RecTaskAlloc(z,F) {
 A := {x |  In(x,p,t,z) ∈ F};
 Q := ∅;
 foreach (PlanTypes P ∈ PlanTypes(z)) {
    InitTaskAlloc(P,F,A);
    do {
        G := NextAlloc(P,F);
        if  (G = FAILURE) return FAILURE;
        zn :=  state  of a in P according to G;
        if (zn  exists ) H := RecTaskAlloc(zn,F∪G);
        else  H:= ∅;
    } while (H = FAILURE);
    Q := Q∪G∪H;
 }
 return  Q;
}
```

Listing 5.4: Recursive Task Allocation Algorithm

be expanded, as the set of agents and the set of tasks is finite, and all valid task allocations satisfy the cardinalities.

Since C is valid, its utility and its heuristic are not smaller than 0. Further, it satisfies all task cardinalities of p and, together with F, models pre- and runtime conditions of p. Thus, it also satisfies all conditions in **AllocGoal**.   □

Note that both Definition 5.19 and Listing 5.2 refer to an agent's capability to prove pre- and runtime conditions by ⊢. An incomplete proving algorithm can cause allocations to be invalid which would be valid otherwise. In practice however, these conditions are often very simple and can be expressed by a few imperative functions mapping onto Boolean values.

Given a task allocation algorithm, a recursive extension is straight forward. Listing 5.4 shows the recursive version. **RecTaskAlloc** allocates agents to all plantypes within the given state z. It does so by recursively calling itself with the state the allocating agent must inhabit if it adopts the resulting allocation.

Proposition 3. *A task allocation done by agent a using **RecTaskAlloc** for state z under assumptions F is recursively valid.*

Proof. Let C be the resulting recursive allocation. Since C is constructed as the union of task allocations by calls to **NextAlloc**, by Proposition 1, it satisfies all pre- and runtime conditions of plans a is allocated to, thus Condition 5.5 of Definition 5.21 holds. Conditions 5.6 and 5.9 also hold due to Proposition 1.

Since **RecTaskAlloc** recursively descends the plan structure using the state zn, which a allocated itself to, Condition 5.7 holds. Condition 5.8 holds since **Expand** only allocates to initial states. Condition 5.10 holds since agents are only allocated if they inhabit the original state z.

Finally, no agent is allocated twice within the same plantype, **InitTaskAlloc** removes all agents which already inhabit the corresponding plantype, and C only contains literals of the form $In(a, p, \tau, z)$ for some agent a, plan p, task τ, and initial state z, hence Condition 5.4 holds.   □

Proposition 4. *A task allocation done by agent a using **RecTaskAlloc** for state z under assumptions F is recursively complete.*

Proof. Let C be the resulting recursive allocation. By Proposition 3, C is recursively valid. Since C is the union of valid allocations produced by **NextAlloc**, and **RecTaskAlloc** returns an allocation only if for all states a is allocated to, the recursive call returns an allocation, and since a is allocated to at most one state per plan, C together with F entail a valid allocation for every plantype in every state a is allocated to. This satisfies Condition 5.11 of Definition 5.22.   □

Proposition 5. *A task allocation done by agent a using **RecTaskAlloc** for state z under assumptions F is recursively perfectly complete if so required.*

Proof. Let C be the resulting recursive allocation. C is recursively complete. Since **Expand** subsequently allocates all agents to tasks if so required, and **AllocGoal** requires the set of unallocated agents to be empty, all solutions to the task allocation problem initialised by **InitTaskAlloc**(P,F,A), allocate all agents in A to a plan in P. Since A contains precisely those agents, which inhabit the parent state, Condition 5.12 and Condition 5.13 hold.   □

Proposition 6. *If a recursively perfectly complete task allocation exists, **RecTaskAlloc** returns one, if so required.*

Proof. A recursively perfectly complete task allocation must allocate agents to all available agents in all plantypes involved according to Condition 5.12 and 5.13. Since **RecTaskAlloc** iterates over all valid task allocations for each plantype (by Proposition 2) until the respective recursive problem is solved, and since all involved conditions and utilities are plan local, the claim holds.   □

Proposition 7. *If a recursively complete task allocation exists, **RecTaskAlloc** returns one, if so allowed.*

Proof. Analogous to the proof for Proposition 6, except that in each single alloca-tion step, **NextAlloc** exhaustively iterates over all possibilities to drop agents, due to **Expand**. If any of them satisfies **AllocGoal**, it is returned eventually. □

Thus, the algorithm in Listing 5.4 constitutes a sound and complete solution to the recursive task allocation problem. However, returning to the restaurant sce-nario in Example 5.1, agreement of the agents involved is still not guaranteed. The issue here is that the plantype holding solely DeliverOrder is not perfectly com-patible with the task DeliverOrder in plan ServeGuests, thus ServeGuests is not perfectly sound. However, the plantype is compatible and hence ServeGuests is sound. Therefore, given the same belief bases the agents a, b and c can agree on a complete allocation, but not on a perfectly complete one.

Definition 5.31. A set of allocations $\mathcal{C}$ *agree*, if $\bigcup \mathcal{C}$ is consistent with Σ_B.

A set of agents A, each with belief base B_i, is in *agreement* about the allocation for a plantype P, if $\bigcup_i \{\text{In}(a, p, \tau, z) \mid \text{In}(a, p, \tau, z) \in B_i \wedge p \in P\}$ is consistent with Σ_B.

Lemma 1. *If all agents $a \in A$ individually compute an allocation for plantype P based on the same assumptions $\mathcal{F}$, they arrive at the same resulting allocation.*

Proof. This follows trivially from the fact that each agent uses the same assump-tions, that Σ_B is common knowledge, and that plans, tasks, and agents are totally ordered by the same ordering. □

Proposition 8. *If plan p is sound, then for all finite sets of agents A, for all states z in p, A is in agreement about the recursively complete allocation RecTaskAlloc(z, B).*

Proof. By Lemma 1, the agents compute the same allocations for each individ-ual plantype, given the same assumptions. Since the initial assumptions B are the same, we only need to show that each agent updates its assumptions in the same way, as it recursively computes the allocation. Since all plans only contain plan local formulae and utilities, individual allocations do not depend on other individ-ual allocations. Further, since p is sound, the only calls to **NextAlloc** which can fail are on the topmost level, i.e., for plantypes in PlanTypes(z). Since all agents try to compute an allocation for all plans in PlanTypes(z), they agree on this level by Lemma 1. □

Proposition 9. *If plan p is perfectly sound, then for all finite sets of agents A, for all states z in p, A is in agreement about the recursively perfectly complete allocation RecTaskAlloc(z, B).*

Proof. Analogously to the proof of agreement for the sound case. □

Of course, these conditions are only relevant when the agents in question have the same belief bases, which is rarely the case in realistic scenarios. Conflicting beliefs can easily lead to conflicting allocations. In Chapter 6, we will show how these cases can be detected and resolved.

With recursive task allocation as an essential tool available, we discuss the rule-based operational semantics of pALICA in the next section. In the following, we denote the result of a task allocation done by agent a for plantype P using agents A in the belief state $\mathcal{F}$ by $\mathrm{TAlloc}(a, P, A | \mathcal{F})$ and the result of a recursive task allocation done by agent a for state z in the belief state $\mathcal{F}$ by $\mathrm{RecTAlloc}(a, z | \mathcal{F})$. Furthermore, we denote a failure in finding a valid task allocation by $\bot$, e.g., if $\mathrm{RecTAlloc}(a, z | \mathcal{F}) = \bot$, then there is no valid recursive task allocation for z given $\mathcal{F}$.

Note that we do not constrain whether the completeness or the perfect completeness property are used to calculate allocations. As will be evident in the next section, both cases can be used with the same operational rules. It depends on the scenario and the plan structure which of the two properties should be used.

5.13 Rules

Transition rules define how an agent's configuration changes during a single computation step. Each rule is guarded by a condition and transforms a given configuration into a new one:

$$\text{RuleName:} \quad \frac{\text{Condition}}{\text{Current Agent Configuration} \longrightarrow \text{New Agent Configuration}}$$

Since multiple rules may be applicable in a given situation, a precedence relationship is introduced. This precedence determines which out of a set of applicable rules is applied. If rule r_1 has a higher precedence than r_2, we write $r_1 > r_2$. The precedence relation over rules, $>$, is transitive and asymmetric.

In the following, a denotes the agent subject to the transition system. A rule is applied only if its condition is satisfied, the agent's configuration, $\mathrm{Conf}(a)$, unifies with the left hand side of the transition and no rule with higher precedence is applicable.

We distinguish two kinds of transition rules, *operational* rules, which describe an agent's normal operation and *repair* rules, which provide means to recover from a failure. Execution of an ALICA program begins with the *Init* rule. From there on, the rules *Trans* and *STrans* describe how an agent reacts to transitions within plans.

Trans captures the case of normal transitions, *STrans* the case of synchronised transitions, which require establishing mutual belief between all involved agents. The application of both rules is followed by the application of rule *Alloc*, which handles recursive task allocation for agents believed to be in the newly entered state. Finally, the rules *BSuccess* and *TSuccess* modify an agent's configuration due to success signals from a lower level (in case of *BSuccess*) and due to reaching a success state within a plan (*TSuccess*). At any point, the rule *Sense* is used to express the incorporation of new sensor information into the belief base.

In principal, these seven rules are sufficient to describe the operative behaviour of the agents involved, unless a failure occurs. Failure handling is done using repair rules. Here, we present a set of ten repair rules, capturing different kinds of reaction to failures. Some of these rules are not applicable in all domains, hence repair rules are somewhat domain dependent. The rule system is meant to be adapted to the specific needs of a domain.

The rule *BAbort* stops a behaviour that is executed if it signals a failure. *BRedo* and *BProp* handle this failure. *BRedo* tries to re-execute a failed behaviour if possible, while *BProp* propagates the failure upwards to the plan in whose context the failed behaviour was executed. *PAbort* acts similar towards plans as *BAbort* does towards behaviours, it stops the failed plan and all plans and behaviours executed in its context. However, it can be overridden by *PRedo* which resets the agent's state within a failed plan if possible. This avoids computational and possibly communication overhead, as the agent continues to work on its task, and does not calculate a new allocation.

In case such a "soft" restart through *PRedo* is impossible and *PAbort* stops a plan, *PReplace* triggers a new task allocation. A new task allocation can also choose an alternative plan from the corresponding plantype. If all other means of failure handling are exhausted, *PProp* propagates a failure upwards to the parent plan. Finally, *PTopFail* captures the case where the top-level plan has failed, and simply triggers a clear initialisation through *Init*.

There are three special rules, *NExpand*, which triggers a failure if a due task allocation cannot be performed, and *Adapt* which is used to periodically check the utility of a task allocation, and which performs a task reallocation if the current utility is deemed to be unsatisfying. *Adapt* thereby allows for highly dynamic changes in the allocation and thus accommodates for swift changes in the environment. Finally, *RoleAlloc* handles role allocation and reallocation. Since reallocations usually happens whenever the team composition or some capabilities change, we regard it as a repair rule as well.

5.13.1 Operational Rules

Operational rules always take precedence over repair rules, following the idea that a failure might become irrelevant by another change in an agent's configuration. Repair rules are geared to preserve a certain status or provide alternatives for a currently pursued, but failed, intention. This also allows one to specify domain-specific handling of failures by referring to a failed sub-plan in a condition of a transition.

The Initialisation Rule

$$Init : \frac{\top}{(B, \emptyset, E, R) \longrightarrow (B + \{In(a, p_0, \tau_0, z_0), Alloc(z_0)\}, \{(p_0, \tau_0, z_0)\}, \emptyset, R)}$$

Intuitively, the Initialisation Rule obligates the agent to start the execution of the plan tree. This is due whenever an agent's plan base is empty, i.e., after start up and whenever an agent's plan base has been emptied completely due to plan failures.

The Sensing Rule

$$Sense : \frac{Sense(\phi)}{(B, \Upsilon, E, R) \longrightarrow ((B + \phi), \Upsilon, E, R)}$$

The sensing rule incorporates sensory information into the belief base. The special predicate $Sense(\phi)$ denotes the sensory information available. We assume that communicated information is incorporated in the same way. One of the key messages exchanged between ALICA agents is the periodic (or semi-periodic) broadcast of the plan-base information together with success information. Such a message encodes beliefs of the form $In(a, p, \tau, z)$ for every triple (p, τ, z) in the plan base of a, and negative beliefs of the form $\neg In(a, p, \tau, z)$ for every triple not in the plan base. Thus the belief base can be updated by removing all beliefs of the form $In(a, p, \tau, z)$ for agent a, and adding those entailed by the received message afterwards. Success information can be treated similarly.

Most sensory information should be handled by a perception component outside of ALICA and correspondingly update the agent's belief state, as described in Section 5.2. We list this rule merely for completeness sake, so that all relevant updates can be described using rules.

The Transition Rule   The Transition Rule controls when and how an agent follows a transition from one state to another.

$$Trans : \frac{B \vdash \phi \wedge (p, \tau, z) \in \Upsilon \wedge (z, z', \phi) \in \mathcal{W} \wedge \neg (\exists s \in \Lambda)(z, z', \phi) \in s}{(B, \Upsilon, E, R) \longrightarrow ((B - \vartheta_b^-) + \vartheta_b^+, (\Upsilon - \vartheta_p^-) \cup \vartheta_p^+, E', R)}$$

where

- $\vartheta_b^- = \{\mathrm{In}(a',p,\tau,z) \mid B \vdash \mathrm{In}(a',p,\tau,z)\}$
  $\cup \{\mathrm{In}(a',p',\tau',z'') \mid a' \in \mathcal{A} \wedge p' \in \mathrm{Plans}^+(z) \wedge z'' \in \mathrm{States}(p')\}$

- $\vartheta_b^+ = \{\mathrm{In}(a',p,\tau,z') \mid \mathrm{In}(a',p,\tau,z) \in \vartheta_b^-\} \cup \{\mathrm{Alloc}(z')\}$

- $\vartheta_p^+ = \{(p,\tau,z')\}$

- $\vartheta_p^- = \{(p,\tau,z)\} \cup \mathrm{Plans}^+(\Upsilon,z)$

- $E' = E - \{(b,z'') \mid (p',\tau',z'') \in \vartheta_p^-\}$

An agent will follow an outgoing transition from state z to z' if it currently resides in z and believes the condition ϕ annotating the transition to hold. Furthermore, this transition must not belong to a synchronisation set. Following a transition entails that the agent stops executing all plans and behaviours that are executed in the context of z, i.e., are in $\mathrm{Plans}^+(\Upsilon,z)$. The addition of $\mathrm{Alloc}(z')$ to the belief base encodes the need for a task allocation with respect to the newly entered state z'. Note that an agent applying this rule also assumes that every other agent currently in z applies it, i.e., believes its precondition. This realises a partial tracking of other agents through the plan tree.

The Synchronised Transition Rule handles a transition within a synchronisation set. Intuitively, a synchronisation models the start of a cooperative act that depends on the involved agents to act in a very small time frame. The upper bound on the size of this time frame depends on the latency and reliability of the communication and the precision with which agents can track their teammates' intentions. In the worst case, the condition guarding the Synchronised Transition Rule cannot be established.

$$STrans: \frac{(\exists A \subseteq \mathcal{A})a \in A \wedge (\exists s \in \Lambda)(z,z',\phi) \in s \wedge \psi}{(B,\Upsilon,E,\mathrm{R}) \longrightarrow ((B - \vartheta_b^-) + \vartheta_b^+, (\Upsilon - \vartheta_p^-) \cup \vartheta_p^+, E',\mathrm{R})}$$

where

- $(p,\tau,z) \in \Upsilon$

- $\psi = (\forall(z'',z''',\phi_i) \in s)(\exists a' \in A, \tau' \in \mathrm{Tasks}(p))B \vdash$
  $\mathrm{MBel}_A(\mathrm{In}(a',p,\tau',z'') \wedge \phi_i)$

- $\vartheta_b^- = \{\mathrm{In}(a',p,\tau',z'') \mid a' \in A \wedge \tau' \in \mathcal{T} \wedge z'' \in \mathrm{States}(p)\} \cup$
  $\{\mathrm{In}(a',p',\tau'',z''') \mid a' \in A \wedge (B \vdash \mathrm{In}(a',p,\tau',z'')) \wedge p' \in \mathrm{Plans}^+(z'')\}$

- $\vartheta_b^+ = \{\text{In}(a', p, \tau', z''') \mid a' \in A \wedge (B \vdash \text{In}(a', p, \tau', z'')) \wedge (z'', z''', \phi) \in s\}$
  $\cup \{\text{Alloc}(z')\}$

- $\vartheta_p^+ = \{(p, \tau, z')\}$

- $\vartheta_p^- = \{(p, \tau, z)\} \cup \text{Plans}^+(\Upsilon, z)$

- $E' = E - \{(b, z) \mid (p, \tau, z) \in \vartheta_p^-\}$

Here, an agent will follow a synchronised transition only if it can identify a group A of agents it is part of, such that a believes that there is mutual belief in A that all relevant conditions ϕ_i hold. Moreover, a has to believe that there is mutual belief in A that all agents in A are in the correct states, that is, that every transition in the synchronisation set will be used by one agent. Hence, there is mutual belief about the individual intentions to progress along the synchronised transitions. If the agent believes this condition to hold, it will act in the same manner as in the case of a normal transition rule. Additionally, it assumes that all participating agents do the same.

Synchronised Transition Rules take precedence over normal transition rules, *STrans > Trans*. This is done following the intuition that a synchronisation guards a part of a plan that is of higher benefit to the team and more difficult to reach. It is easy to see that the condition of *Trans* would subsume the condition of *STrans* if it were not for the exclusion of synchronisations $(\neg(\exists s \in \Lambda)(z, z', \phi) \in s)$.

Synchronisations realise, as the name suggests, a tight coupling between agents. The mutual belief is established via communication (see Section 7.5 for details). In some cases, this additional communication is counter productive, depending on the quality and load of the communication medium. In such a case, a weaker form of synchronisation can be used, based on ordinary transitions. Suppose two transitions (z_1, z_2, ϕ) and (z_3, z_4, ψ) are to be synchronised. Assuming neither z_2 nor z_4 are initial states and no other transition leads to z_2 or z_4, a behaviour similar to synchronisations, but less strict can be achieved by replacing ϕ with $\phi \wedge \psi \vee (\exists a, p, \tau) \text{In}(a, p, \tau, z_4)$ and ψ with $\phi \wedge \psi \vee (\exists a, p, \tau) \text{In}(a, p, \tau, z_2)$. Thus, the agents will follow one another. This weaker kind of synchronisation requires only a single message, but is more susceptible to packet loss or delay.

The Allocation Rule   The Allocation Rule takes over where the transition rules above left an agent. It causes a task allocation to be performed, usually in a state just entered.

$$Alloc: \frac{(B \vdash \text{Alloc}(z) \wedge \text{In}(a, p, \tau, z)) \wedge \text{RecTAlloc}(a, z|B) \neq \bot}{(B, \Upsilon, E, R) \longrightarrow ((B - \{\text{Alloc}(z)\}) + \vartheta_b^+, \Upsilon \cup \vartheta_p^+, E \cup \vartheta_e^+, R)}$$

where

- $\vartheta_b^+ = \mathrm{RecTAlloc}(a,z|B)$
- $\vartheta_p^+ = \{(p',\tau',z') \mid \mathrm{In}(a,p',\tau',z') \in \vartheta_b^+\}$
- $\vartheta_e^+ = \{(b,z') \mid ((p,\tau,z') \in \vartheta_p^+ \lor z' = z) \land b \in \mathrm{Behaviours}(z')\}$

That is, if a believes a task allocation with respect to state z is needed and possible, it will update its believe base with the computed allocation $\mathrm{RecTAlloc}(a,z|B)$, as introduced in Section 5.12, insert all plans it is involved in into its plan base, and start to execute all behaviours relevant to the added plan-task-state-triples.

Since the result of an allocation is only relevant until the agent leaves the corresponding state, Transition Rules have a higher precedence than the Allocation Rule, *Trans* > *Alloc*.

The Behaviour Success Rule   In ALICA, behaviours are atomic actions, which can fail or succeed at any point in time during their execution. Such a termination is reflected by the atoms $\mathrm{Success}(b)$ and $\mathrm{Fail}(b)$, where b is the behaviour in question. Since b is embedded in a canonical behaviour plan or a plan with similar properties, the agent will subsequently enter the success state of that plan by means of the Transition Rule.

$$BSuccess : \frac{(b,z) \in E \land B \vdash \mathrm{Success}(b)}{(B,\Upsilon,E,\mathrm{R}) \longrightarrow (B,\Upsilon,E - \{(b,z)\},\mathrm{R})}$$

where z is the state which acts as the context of b. If a running behaviour succeeds, it is stopped. By the Planning Axiom, its postcondition should hold as well. Since this rule processes a signal and updates the execution set accordingly, it takes precedence over the previous *Trans* and *Alloc* rules which modify the plan base.

Task Success Rule   An agent succeeds in completing a task within a plan, if and only if it reaches a state $z \in \mathrm{Success}(p)$.

$$TSuccess : \frac{(p,\tau,z) \in \Upsilon \land z \in \mathrm{Success}(p)}{(B,\Upsilon,E,\mathrm{R}) \longrightarrow (B + \vartheta_b^+,\Upsilon,E,\mathrm{R})}$$

where $\vartheta_b^+ = \{\mathrm{Succeeded}(a,p,\tau),\mathrm{Post}(z)\}$.

This rule updates the belief base with the postcondition attached to the terminal state reached. Moreover, the successful completion of task τ in p is recorded as well.

This update might lead to an inconsistent allocation, if $\mathrm{TeamIn}(p)$ no longer holds due to the changed cardinalities $\xi_t(p, \tau, B + \vartheta_b^+)$. A subsequent task reallocation by the Adaptation Rule will reallocate agent a correspondingly, unless a transition on the parent level reacts on the successful completion of the task. The success needs to be communicated to other agents working on p, otherwise, it might be the case that they abort the plan due to an insufficient number of agents working on it, once a itself reacts to the success (see Example 5.3). Thus, an agent is committed to informing its team about the achieved goal represented by $\mathrm{Succeeded}(a, p, \tau)$, similar to a joint intention.

We give precedence to the success of lower level behaviours, but still treat this success rule with a higher priority than the Transition Rules, *BSuccess* > *TSuccess* > *STrans*. This way, a subsequent check of a transition rule can react to the success of the task.

Example 5.3 (Asynchronous Plan Success). *This example shows the case that one task of a plan is finished before another task of that plan. As aforementioned the successful completion of a task is represented in the belief base and communicated to other agents. Figure 5.4 illustrates such a case. Initially the two agents a and b are allocated to the tasks τ_1 and τ_2 of plan p_1. Both tasks have an associated cardinality of 1..1, which means that exactly one agent must be allocated to each task for the execution of the plan. Both agents start with the initial state of their task within the plan. Agent a executes plantype P_2 and agent b plantype P_3. After some time, agent a progresses to the next state s_2. Agent b leaves state s_3 following the transition to the success state s_4. Now agent b leaves the plan and therefore it is not longer committed to task τ_2. If the successful completion of task τ_2 were not recorded, the plan would be aborted as the cardinalities of the task are not satisfied. See also the definition of* TeamIn *(Definition 5.12). With the success being communicated, assuming τ_2 is the only task in* $\mathrm{Required}(p_1)$*, agent a would be free to leave the plan as well, as p_1 is successfully completed.*

5.13.2 Repair Rules

Typical BDI languages feature techniques handling failures that occur, e.g., due to unexpected changes in the dynamic environment agents act in. Classic BDI languages also distinguish between plan failures and goal failures. The first can be due to some side constraint being violated, the latter due to the goal itself becoming impossible to achieve. Refer for instance to Sardina and Padgham [141].

While goals are represented as appropriate postconditions, ALICA does not come with any reasoning capability to infer whether or not a goal is still achievable

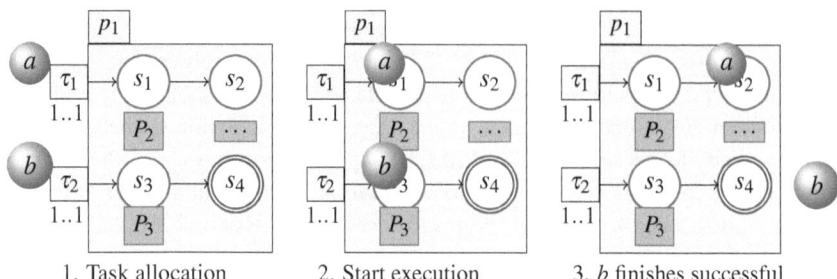

Figure 5.4: Example: Asynchronous Plan Success

given only domain knowledge. Instead, the failure to achieve a goal or the belief in unachievable goals is represented using plan elements such as runtime conditions and failure states.

The corresponding notion of plan failures can happen frequently, depending on the scenario. For instance in the dynamic robotic soccer domain, each robot has to make assumptions about the status of its teammates, which can prove to be wrong. Repair rules are special transition rules which are meant to recover from such failures. There are different ways to handle failures and a specific way can depend on the domain. A failed plan can be retried, replaced by an alternative, or the failure can be propagated up the plan tree. In some languages, such as AgentSpeak [129], a failure can even raise a specific goal, triggering custom plans meant to handle precisely the occurred failure.

ALICA features a similar way to deal with plan failures explicitly. A failed plan or behaviour is recognised and causes a corresponding believe to be inserted in the belief base (Handle$_f(p)$ for a plan, and Handle$_f(b,z)$ for a behaviour, respectively, as described in Section 5.2.2). Since the Transition Rule takes precedence over repair rules, an explicit mechanism can be modelled via a transition. Otherwise, default handling takes place.

The Behaviour Abortion Rule   A behaviour is aborted if it signals a failure. Thus, it is removed from the execution set.

$$BAbort: \frac{(b,z) \in E \wedge B \vdash \text{Fail}(b)}{(B,\Upsilon,E,\text{R}) \longrightarrow (B',\Upsilon,(E - \{(b,z)\}),\text{R})}$$

where

- $B' = (B - \{(\forall i)\,\text{Failed}(b,z,i),\text{Fail}(b)\})$
  $+ \{\text{Failed}(b,z,j),\text{Handle}_f(b,z)\},$

- $j = \begin{cases} i+1 & \text{if Failed}(b,z,i) \in B, \\ 1 & \text{otherwise.} \end{cases}$

The belief $\text{Failed}(b,z,i)$ keeps track of how many times a behaviour had to be aborted. This allows the consecutive application of different failure recovery rules. Note that if the corresponding state z is left, e.g., through a transition, this belief is dropped to keep the belief base consistent with Σ_b (see Definition 5.6).

The Behaviour Repair Rules act as a default mechanisms to handle behaviour failure.

$$BRedo : \frac{B \vdash \text{In}(a,p,\tau,z) \wedge \text{Failed}(b,z,1) \wedge \text{Handle}_f(b,z)}{(B,\Upsilon,E,R) \longrightarrow (B',\Upsilon,E',R)}$$

where

- $E' = E \cup \{(b,z)\}$

- $B' = B - \{\text{Handle}_f(b,z)\}$

$$BProp : \frac{B \vdash \text{In}(a,p,\tau,z) \wedge (\exists i)\,\text{Handle}_f(b,z) \wedge \text{Failed}(b,z,i) \wedge i > 1}{(B,\Upsilon,E,R) \longrightarrow (B',\Upsilon,E,R)}$$

where

- $B' = (B - \{\text{Handle}_f(b,z),\text{Failed}(b,z,i),(\forall k)\,\text{Failed}(p,k)\})$
  $+ \{\text{Handle}_f(p),\text{Failed}(p,j)\}$

- $j = \begin{cases} j'+1 & \text{if Failed}(p,j') \in B \\ 1 & \text{otherwise} \end{cases}$

BRedo restarts a failed behaviour if possible, and *BProp* propagates the failure upwards to the containing plan if restarting has already been tried. The universal quantification in $(\forall k)\,\text{Failed}(p,k)$ ensures that $\text{Failed}(p,k)$ is removed regardless of the current value of k. It is the case that at most one instance of $\text{Failed}(p,k)$ holds per plan p. Note that the applicability of these rules is subject to the concrete domain. For instance, retrying a failed behaviour might not make sense at all in certain scenarios. In others, a behaviour's success can be associated with a known probability distribution, in which case the utility of a retry can be estimated. Therefore, it is sensible to customise these rules with respect to the domain.

The Plan Abortion Rule   is quite similar to the Behaviour Abortion Rule:

$$PAbort : \frac{(p,\tau,z) \in \Upsilon \wedge (z \in \text{Fail}(p) \vee B \vdash \neg \text{Run}(p) \vee \neg \text{TeamIn}(p))}{(B,\Upsilon,E,\text{R}) \longrightarrow ((B - \vartheta_b^-) + \vartheta_b^+, \Upsilon - \vartheta_p^-, E - \vartheta_e^-, \text{R})}$$

- $\vartheta_b^- = \{\text{In}(a',p,\tau',z') \mid a' \in \mathcal{A}, \tau' \in \mathcal{T}, z' \in \text{States}(p)\}$
  $\cup \{\text{In}(a',p',\tau',z') \mid (\exists z'' \in \text{States}(p))p' \in \text{Plans}^+(z'')\}$
  $\cup \{(\forall k)\,\text{Failed}(p,k)\}$

- $\vartheta_b^+ = \{\text{Handle}_f(p),\text{Failed}(p,j)\}$

- $\vartheta_p^- = \{(p,\tau,z)\} \cup \text{Plans}^+(\Upsilon,z)$

- $\vartheta_e^- = \{(b,z) \mid (p,\tau,z) \in \vartheta_p^-\}$

- $j = \begin{cases} j'+1 & \text{if Failed}(p,j') \in B \\ 1 & \text{otherwise} \end{cases}$

This rule aborts a plan if the corresponding runtime condition is violated, a failure state is reached or if the agent believes that the team no longer executes the plan. Additionally, the rule aborts all plans and behaviours executed in the context of the current state z, and assumes that all agents involved in executing p detect the plan failure and apply the same rule accordingly. If another participating agent does not detect the failure, this leads to conflicting belief states, where some agents still believe that the plan can be executed, while others do not. Resolving this state requires the agent to inform its teammates about the plan failure. A periodic broadcast of the plan base contains sufficient information to deduce the plan failure, or compensate for the missing agent. In Section 7.5, we will discuss these messages in more detail.

The Plan Repair Rules   implement default ways to handle failed plans. Intuitively, a plan can be restarted, replaced by an alternative or the failure can be propagated upwards.

$$PTopFail : \frac{B \vdash \text{Handle}_f(p_0)}{(B,\Upsilon,E,\text{R}) \longrightarrow (B - \{(\forall p,\tau,z)\,\text{In}(a,p,\tau,z),\text{Handle}_f(p_0)\},\emptyset,\emptyset,\text{R})}$$

PTopFail handles failures of the top-level plan by resetting the agent configuration, thus triggering *Init* again. The only way to handle a failure at this level is to retry the whole program.

$$PRedo : \frac{(p,\tau,z) \in P \wedge z \in \text{Fail}(p) \wedge (B \vdash \neg(\exists x)\,\text{Failed}(p,x)) \wedge B' \vdash \psi}{(B,\Upsilon,E,R) \longrightarrow (B',(\Upsilon - \vartheta_p^-) \cup \{(p,\tau,z')\},E',R)}$$

where

- $B' = (B - \vartheta_b^-) + \{In(a,p,\tau,z'), \text{Alloc}(z'), \text{Failed}(p,1)\}$

- $\vartheta_b^- = \{In(a,p,\tau,z)\} \cup \{In(a,p',\tau',z'') \mid \tau' \in \mathcal{T} \wedge p' \in \text{Plans}^+(\Upsilon,z)\}$

- $z' = \text{Init}(p,\tau)$

- $\psi = \text{TeamIn}(p) \wedge \text{Pre}(p) \wedge \text{Run}(p)$

- $\vartheta_p^- = \{(p,\tau,z)\} \cup \text{Plans}^+(\Upsilon,z)$

- $E' = E - \{(b,z) \mid (p,\tau,z) \in \vartheta_p^-\}$

By applying *PRedo*, an agent retries to fulfil its task within a plan p, if it has reached a failure state and believes its team is still working on p and that the pre-conditions and runtime conditions are still met. Note that evaluating the condition of this rule requires the agent to hypothesis B' before applying *PRedo*. *PRedo* takes precedence over *PAbort*, *PRedo > PAbort*, so a less extensive failure handling is tried first. Subsequent failures of the same plan will not be handled by *PRedo* due to the insertion of Failed$(p,1)$ into the belief base. Since there is only one way to handle failure of the top-level plan, *PTopFail > PRedo*.

$$PReplace : \frac{B \vdash \text{Handle}_f(p) \wedge \text{Failed}(p,1)}{(B,\Upsilon,E,R) \longrightarrow (B',\Upsilon,E,R)}$$

where

- $B' = (B - \{\text{Handle}_f(p)\}) + \{\text{Alloc}(z)\}$

- $p \in \text{Plans}(z)$

PReplace handles a failure by triggering a new task allocation for the state in which p is executed.

$$PProp : \frac{(B \vdash \text{Handle}_f(p) \wedge \text{Failed}(p,2)) \wedge (p',\tau,z) \in \Upsilon}{(B,\Upsilon,E,\theta,R) \longrightarrow ((B - \vartheta_b^-) + \vartheta_b^+, \Upsilon - \vartheta_p^-, E - \vartheta_e^-, \theta, R)}$$

where

- $p \in \text{Plans}(z)$

- $\vartheta_b^- = \{\text{In}(a', p', \tau', z) \mid a' \in \mathcal{A} \wedge \tau' \in \mathcal{T}\}$
  $\cup \{\text{In}(a', p'', \tau', z') \mid (\exists z'' \in \text{States}(p')) p'' \in \text{Plans}^+(z'')\}$
  $\cup \{(\forall k)\, \text{Failed}(p', k)\}$

- $\vartheta_b^+ = \{\text{Handle}_f(p'), \text{Failed}(p', j)\}$

- $\vartheta_p^- = \{(p', \tau, z)\} \cup \text{Plans}^+(\Upsilon, z)$

- $\vartheta_e^- = \{(b, z) \mid (p', \tau, z) \in \vartheta_p^-\}$

- $j = \begin{cases} j' + 1 & \text{if } \text{Failed}(p', j') \in B \\ 1 & \text{otherwise} \end{cases}$

The last option for an agent confronted with a plan failure is to propagate the failure upwards, which is done here by aborting the parent plan and triggering failure handling rules for it. Since a failure should be handled at the lowest level possible, *PReplace > PProp*.

The Allocation Failure Rule handles the case where a task allocation cannot assign any agent to a plan, for instance if a precondition cannot be met. If an allocation for state z fails, a failure for the corresponding plan p is raised.

$$NExpand : \frac{(B \vdash \text{Alloc}(z)) \wedge \text{RecTAlloc}(a, z \mid B) = \bot \wedge (p, \tau, z) \in \Upsilon}{(B, \Upsilon, E, R) \longrightarrow ((B - \vartheta_b^-) + \vartheta_b^+, \Upsilon - \vartheta_p^-, E - \vartheta_e^-, R)}$$

where

- $\vartheta_b^- = \{\text{Alloc}(z)\} \cup \{\text{In}(a', p, \tau', z) \mid a' \in \mathcal{A} \wedge \tau' \in \mathcal{T}\}$
  $\cup \{\text{In}(a', p', \tau', z') \mid (\exists z'' \in \text{States}(p)) p' \in \text{Plans}^+(z'')\}$
  $\cup \{(\forall k)\, \text{Failed}(p, k)\}$

- $\vartheta_b^+ = \{\text{Handle}_f(p), \text{Failed}(p, j)\}$

- $\vartheta_p^- = \{(p, \tau, z)\} \cup \text{Plans}^+(\Upsilon, z)$

- $\vartheta_e^- = \{(b, z) \mid (p, \tau, z) \in \vartheta_p^-\}$

- $j = \begin{cases} j' + 1 & \text{if } \text{Failed}(p, j') \in B \\ 1 & \text{otherwise} \end{cases}$

The Adaptation Rule   treats the case where a plan has not failed but has a comparatively low utility evaluation. In this case, an agent can trigger a new task allocation if it believes there exists an allocation which is more suitable to the current situation. This yields a task reallocation mechanism similar to the local decisions discussed by Nair et al. [104]. However, task reallocation in this way is seamlessly integrated into the language and happens on-the-fly without further need for expensive deliberation.

An allocation is replaced by a new one in case the utility of the new one is deemed higher. In order to stabilise decisions, we use a plan-specific threshold value, $t(p)$, that must be exceeded by the difference in order for the reallocation to take place. Additionally, we use a similarity measure $\mathrm{Sim}(p,B,B')$, weighted by the plan-specific factor $w_s(p)$, such that reallocations that are very different must have a correspondingly higher utility.

$$Adapt: \frac{\mathcal{U}(p_n)(B'') - w_s(p_c)\,\mathrm{Sim}(p_c,B'',B) > \mathcal{U}(p_c)(B) + t(p_c)}{(B,\Upsilon,E,\mathrm{R}) \longrightarrow (B_n,\Upsilon_n,E_n,\mathrm{R})}$$

where:

Agents A currently inhabit state z:

$$A = \{a' \mid \mathrm{In}(a',p,\tau,z)\}$$

Agent a is currently executing p_c in the context of z or is passively participating in p_c:

$$a \in A \wedge (B \vdash \mathrm{TeamIn}(A,p_c)) \wedge p_c \in \mathrm{Plans}(z)$$

The plans p_c and p_n belong to the same plantype:

$$p_c \in P \wedge p_n \in P$$

$\vec{P_c}$ are the plans in the sub-branch subject to reallocation:

$$\vec{P_c} = \{p_c\} \cup \{p' \mid (\exists z' \in \mathrm{States}(p_c))p' \in \mathrm{Plans}^+(z')\}$$

B' is the belief base without any assumptions about p_c and its sub-plans:

$$B' = (B - \{\mathrm{In}(a',p',\tau',z') \mid a' \in A \wedge p' \in \vec{P_c}\}$$

B'' is the result of a task allocation for plantype P combined with the assumptions B':

$$B'' = B' + \mathrm{TAlloc}(a,P,A|B')$$

p_n is the plan selected by the valid task allocation:

$$\mathrm{TAlloc}(a,P,A|B') \neq \bot \wedge B'' \vdash \mathrm{TeamIn}(A,p_n)$$

Δ_A^- denotes the reallocated agents:

$$\Delta_A^- = \{a' \mid \mathrm{In}(a',p_c,\tau',z') \in B \wedge \neg(\exists z'')\,\mathrm{In}(a',p_c,\tau',z'') \in \mathrm{TAlloc}(a,P,A|B')\}$$

B_r are the beliefs regarding the sub-branch that remain unchanged:

$$B_r = \{\mathrm{In}(a',p',\tau',z') \mid a' \in A \wedge a' \notin \Delta_A^- \wedge \mathrm{In}(a',p',\tau',z') \in B \wedge p' \in \vec{P}_c\}$$

B''' contains only the beliefs that do not conflict with B_r:

$$B''' = B'' - \{\mathrm{In}(a',p',\tau',z') \mid (\exists z'')\,\mathrm{In}(a',p',\tau',z'') \in B_r\}$$

B_n is the new belief base, containing all unchanged beliefs, the new task allocation, and $\mathrm{Alloc}(z_n)$ in case a beliefs at least one agent was reallocated to z_n, i.e., the state a inhabits after reallocation:

$$B_n = \begin{cases} B''' \cup B_r \cup \{\mathrm{Alloc}(z_n)\} & \text{if } \phi \\ B''' \cup B_r & \text{otherwise} \end{cases}$$

where

$$\phi = \mathrm{In}(a,p_n,\tau_n,z_n) \in (B''' + B_r) \wedge (\exists a')\,\mathrm{In}(a',p_n,\tau_n,z_n) \in B'''$$

The plan base and the execution set are limited to the new beliefs:

$$\Upsilon' = \Upsilon - \{(p',\tau',z') \mid \mathrm{In}(a,p',\tau',z') \notin B_n)$$
$$E_n = E - \{(b,z') \mid (p',\tau',z') \notin \Upsilon')$$

If the agent entered a new state as a consequence of this rule, this is reflected in the new plan base:

$$\Upsilon_n = \Upsilon' \cup \{(p_n,\tau_n,z_n) \mid \mathrm{In}(a,p_n,\tau_n,z_n) \in B_n\}$$

$\text{Sim}(p, \mathcal{F}, \mathcal{G})$ is the similarity measure:

$$\text{Sim}(p, \mathcal{F}, \mathcal{G}) = 1 - \frac{|\{a \mid \text{In}(a, p, \tau, z) \in \mathcal{F} \wedge (\exists z') \text{In}(a, p, \tau, z') \in \mathcal{G}\}|}{\max(1, |\{a \mid (\exists p, \tau, z) \text{In}(a, p, \tau, z) \in \mathcal{G}\}|)}$$

That is, if a new task allocation for a plantype P has a utility higher than the current allocation with respect to the situation at hand, the agent will adopt the new allocation. The threshold value $t(p)$ limits the applicability of this rule. $t(p)$ as well as $w_s(p)$ are specific to each plan p. Note that p_n and p_c can refer to the same plan or to two different plans in the same plantype P. Thereby, this rule allows the agents to switch from one alternative solution to the other in case the former becomes infeasible or is deemed comparatively hard to achieve. Similar to the Transition Rule, the current allocation is removed from belief base, plan base, and execution set. The new allocation is adopted directly, similar to the Task Allocation Rule. Since this rule also implements a soft repair mechanism in case a runtime condition is violated, it has a higher precedence than all repair rules.

The definition of the belief update captures the notion of minimal change, i.e., agents that do not need to change their tasks do not change their states either. Should the local agent be reallocated by *Adapt*, or believes that another agent is reallocated to the state it inhabits, the belief $\text{Alloc}(z_n)$ is inserted into the belief base, triggering a subsequent recursive task allocation.

The Adaptation Rule enables a team to react swiftly to changing situations. Due to the threshold value and the similarity measure, oscillation can be avoided. However, their value depend on the utility functions used and are therefore domain dependent. Further, both threshold and similarity weight can be used to emphasis dynamic adaptation or forbid it altogether for each plan separately. In general, as an agent progresses towards successful completion of its task, the overall utility should increase. One way to achieve this, given reward functions for successful completion, is using utilities which obey the Bellman equations [8].

The Role Allocation Rule   treats the case where a role reallocation is necessary. The reallocation is computed by the formation F (see Definition 5.15), and adapted by the agent in case any agent is allocated differently than before.

$$RoleAlloc : \frac{\{\text{HasRole}(a', r) \mid \text{HasRole}(a', r) \in B\} \neq F(\{a' \mid \text{In}(a', p_0, \tau_0, z_0) \in B\})}{(B, \Upsilon, E, R) \longrightarrow ((B' - \vartheta_b^-) + \vartheta_b^+, \Upsilon, E, R')}$$

where

- $R' = \{\text{HasRole}(a, r) \mid \text{HasRole}(a, r) \in F(\{a' \mid \text{In}(a', p_0, \tau_0, z_0) \in B\})$

- $\vartheta_b^- = \{(\forall a, r) \, \mathrm{HasRole}(a, r)\}$
- $\vartheta_b^+ = \mathrm{F}(\{a' \mid \mathrm{In}(a', p_0, \tau_0, z_0) \in B\})$

As stated in Section 5.6, role reallocation is only needed when the team composition changes, either due to agents leaving or joining the team, or due to a change in an agent's capabilities. For sake of simplicity, the corresponding triggers are not represented explicitly in the rule.

The complete team is represented in an agent's belief base by the set of agents that inhabit the solitary state in the top-level plan, $\{a' \mid \mathrm{In}(a', p_0, \tau_0, z_0)\}$. Thus, it is sufficient to check the applicability of this rule when this set changes or the agent is notified of a change in any participating agent's capabilities.

In summary, the presented repair rules form a flexible and customisable system to react on failures, allowing for robust adaptation of the team's behaviour to changing situations. Table 5.1 gives a brief overview of the introduced rules in an order compatible with the partial precedence order established between rules.

The represented rules are not a minimal set, in the sense that smaller rule sets are sufficient to execute an ALICA program. For instance, the rules *PRedo* and *BRedo* are not necessary, or the rules *Alloc* and *Adapt* could be merged into a single rule. However, the presented set of rules is easier to modify than a minimal set. For example, in domains where dynamic adaptation is not needed or even counter-productive, *Adapt* can simply be removed.

5.14 Agent Configuration Consistency

In this section, we verify the operational semantics with respect to the plan base axioms introduced in Definition 5.4. Additionally, we show that the plan base of an agent always forms a tree, that no behaviour is left orphaned, i.e., the context of a behaviour in execution is always referred to in the plan base, and that an agent always beliefs in what it is doing. This last property is a requirement for a belief base according to Definition 5.8.

According to Definition 5.8 a belief base is required to be consistent with the common knowledge and the plan base.

Proposition 10. *If* $\mathrm{Conf}(a) = (B, \Upsilon, E, R)$ *then* $(p, \tau, z) \in \Upsilon$ *if and only if* $B \vdash \mathrm{In}(a, p, \tau, z)$.

Proof. The initial belief base does not contain any beliefs of the form $\mathrm{In}(a, p, \tau, z)$ and the initial plan base is empty by Definition 5.2. Moreover, every rule that modifies the plan base makes an equivalent update to the belief base. $\qquad \square$

Init	Initialises or reinitialises an agent's configuration.
RoleAlloc	Updates an agent's role assumptions according to a newly computed role allocation.
Sense	Incorporates new sensory information into the belief base.
BSuccess	Reacts upon the success of a behaviour.
TSuccess	Reacts upon the success of a task.
STrans	Moves an agent along a synchronised transition after an appropriate mutual beliefs was established.
Trans	Moves an agent from one state to another.
Alloc	Implements recursive allocation within a state the agent entered.
Adapt	Replaces the current task allocation with a new one if the difference in the corresponding utility values is significant.
BAbort	Aborts the execution of a failed behaviour and raises a corresponding flag.
BRedo	Retries a failed behaviour if possible.
PTopFail	Handles the case where the top-level plan, p_0, failed.
PRedo	Restarts the execution of a task within a failed plan if possible.
PAbort	Aborts a failed plan.
PReplace	Replaces a failed plan by triggering a new task allocation in its parent state.
PProp	Propagates the failure of a plan to its parent plan.
NExpand	Promotes the inability to calculate a valid task allocation to a plan failure.

Table 5.1: Operational Rules of pALICA from Highest to Lowest Precedence

Thus, the belief base is consistent with the plan base and an agent always believes what it intentionally is doing. Consistency with respect to the common knowledge is maintained by the belief update semantics as discussed in Section 5.2.3.

Proposition 11. *If a program is well-formed, all plan bases that can occur during runtime contain at most one triple (p, τ, z) per plan p.*

Proof. In the following, we call a plan base which contains at most one triple per plan consistent. The initial plan base is empty, and therefore fulfils the property. Plan bases are modified through the application of rules. The rule *Init* adds a triple to the empty plan base, thereby producing a consistent plan base.

The rule *Sense, TSuccess, BAbort, BRedo, BProp, BSuccess, PReplace*, and *RoleAlloc* do not modify the plan base. The rules *PAbort, PTopFail, PProp*, and *NExpand* only remove triples from the plan base, thus maintain consistency. The rule *PRedo* adds a single triple (p, τ, z') to the plan base, but also removes (p, τ, z), maintaining consistency.

The rules *Trans* and *STrans* remove the triple which refers to the plan p and state z in which the transition occurs, and all triples belonging to the branch $\text{Plans}^+(z)$. They only add a single new triple, which refers to p. Thus if the input plan base was consistent, both rules produce consistent plan bases.

The rule *Alloc* employs the recursive task allocation and adds the resulting triples to the plan base. Since the program is a tree, the recursive task allocation algorithm allocates an agent at most once to each plan (by Proposition 3), and it takes the current beliefs into account, which are consistent with the input plan base by Proposition 10, all plan bases produced by the rule *Alloc* are consistent if the corresponding input plan base was consistent.

The rule *Adapt* removes all beliefs referring to a sub-branch in case the local agent is reallocated. Consequently, it also removes the corresponding triples from the plan base. The only triple that can be added refers to the newly entered state, since $\text{TAlloc}(a, A, P|\mathcal{F})$ only results in valid allocations with respect to P. If this tuple is added, any tuple referring to the formerly executed plan is removed. Hence, the claim holds for *Adapt* as well.   □

Proposition 11 is a requirement, but not a sufficient condition that the plan base forms a tree at all times.

Proposition 12. *For any tuple* (p, τ, z) *in a plan base* Υ

- $p = p_0$ *or*

- *another tuple* (p', τ', z') *occurs in* Υ *such that* $p \in \text{Plans}(z')$.

Proof. In the following, we call a plan base that fulfils the requirement consistent. The initial plan base is empty, thus the initial plan base is consistent. The rule *Init* adds the tuple (p_0, τ_0, z_0) to the empty plan base, hence the claim holds as well. The rules *Sense, TSuccess, BAbort, BRedo, BProp, BSuccess, PReplace*, and *RoleAlloc* do not modify the plan base. The rules *Trans* and *STrans* remove all triples, which are executed in the context of the state left by the agent, and therefore maintain the claim.

The rule *Alloc* adds a complete branch to the plan base, since RecTAlloc produces only recursively valid task allocations. *Adapt* reallocates in the plantype P. It removes a complete branches from the plan base and only adds a single tuple referring to the plan p_n. Since p_n belongs to the same plantype as the replaced plan p_c, and the tuple (p, τ, z) remains untouched, the resulting plan base stays consistent.

The rules *PAbort*, *PProp*, and *NExpand* remove all plans that are executed in the context of the aborted plan. *PTopFail* results in an empty plan base. *PRedo* removes the triple (p, τ, z) and all tuples in the context of z, and only adds $(p, \tau, \mathrm{Init}(p, \tau))$, thus if the input plan base was consistent the resulting plan base is consistent as well. $\qquad \square$

Theorem 5.1. *If the ALICA program is well-formed, all plan bases that can occur during runtime form a tree.*

Proof. The claim holds due to Proposition 11 and Proposition 12, and due to the well-formedness of the program, which requires the plan-tree to be a tree. $\qquad \square$

The property that a plan base always forms a tree is quite important, it simplifies reasoning and allows implementations to use appropriate data structures, allowing for efficient traversal and rule application.

Given the tree-shape of the plan base, we can now guarantee that the plan base is always consistent with the plan base axioms (see Definition 5.4):

Theorem 5.2. *If the ALICA program is well-formed, all plan bases that can occur during runtime are consistent with the plan base axioms Σ_p.*

Proof. Consistency with Axiom 5.1 is implied by Proposition 11. Axiom 5.2 is satisfied since:

- p_0 cannot belong to a plantype in the program tree. Otherwise, the program would not form a tree. Hence, *Init* maintains consistency.

- RecTAlloc allocates only to one plan per plantype since the program forms a tree.

- *Adapt* removes the triple referring to the formerly executed plan if it adds one.

- No other rule starts to execute a new plan, i.e., adds a tuple (p, τ, z) such that no tuple (p, τ', z') was in the input plan base.

Consistency with Axiom 5.3 is maintained since:

- The empty plan base is consistent with the axiom.

- The tuple (p_0, τ_0, z_0) satisfies the conditions of the axiom.

- *Trans* and *STrans* remove the tuple (p, τ, z) and add the tuple (p, τ, z') only if a transition (z, z', ϕ) exists for some ϕ, thus both maintain consistency.

- Task allocation allocates only to tasks of the respective plan, i.e., which are an element of Tasks(p), and allocates only to initial staes. Hence *Adapt* and *Alloc* maintain consistency.

- No other rule changes the task of an agent or adds a tuple other than $(p, \tau, \text{Init}(p, \tau))$ for some plan p and some task τ such that (p, τ, z) was in the input plan base for some state z.

$\square$

Besides consistency of the plan base, we also want to guarantee that a robot only does what it intends. In other words, it should only execute behaviours that reflect the procedural intentions represented by the plan base.

Proposition 13. *No behaviour is left orphaned. For any agent configuration* (B, Υ, E, R) *that can occur during runtime:*

$$(\forall (b, z) \in E)(\exists p, \tau)(p, \tau, z) \in \Upsilon$$

Proof. Behaviours are added to the execution set by the rules *Alloc* and *BRedo*. *BRedo* requires the agent to inhabit the corresponding state by $\text{In}(a, p, \tau, z)$, and *Alloc* only adds tuples (b, z) such that a corresponding triple (p, τ, z) is added to the plan base at the same time.

Triples are removed from the plan base by the rules *Trans*, *STrans*, *PAbort*, *PTopFail*, *PRedo*, *PProp*, *NExpand*, and *Adapt*. All these rules remove all tuples from the execution set which refer to states of triples removed from the plan base, except *PTopFail*, which results in an empty execution set. $\square$

Consequently, if a behaviour is executed in the context of state z, the agent also believes that it inhabits z.

5.15 Summary

This chapter presented the formal semantics of the language pALICA. We discussed an underlying agent model, which is represented in the semantics by an agent configuration, consisting of a belief base, a plan base, an execution set, and a role set. The belief base captures the current information an agent has about its environment and its team. An agent's internal state with respect to the executed

ALICA program is represented by the plan base, which constitutes a procedural description of the agent's intentions. The execution set consists of the behaviours the agent currently executes, and thus forms the link to lower-level actuator components. Finally, the role set contains all roles an agent assumes within the team.

The main part of the semantics is given by a transitional rule system, which describes in detail how and when the agent configuration is updated. The set of rules consists of operational rules, that guide the normal operation of an agent, and repair rules, which formulate reactions to failures and unexpected changes in the environment. For solving the task allocation problem, a recursive algorithm is presented that allocates agents to tasks at each level in the plan hierarchy. Following the locality and autonomy principles, each agent computes this allocation locally for the plans it participates in. This approach requires a soundness property that given ALICA programs must fulfil in order for the agents to compute compatible allocations. Depending on the degree of coupling between the plans, two different requirements are identified and discussed. Finally, we showed that the internal state of each agent always forms a tree and satisfies necessary consistency axioms, if the program is well-formed according to the previous chapter. This property can be exploited by implementations in order to gain efficiency.

6 Conflict Detection and Resolution

Conflicts in teamwork manifest themselves in ALICA in conflicting task allocations. Each agent maintains a set of beliefs for each plan it participates in, which represents the assignment of agents to tasks within that plan. As discussed in Section 5.8, a conflict with respect to these assignments can be caused by different belief bases in the team, the execution of unsound plans, or the execution of plans which are not perfectly sound if idling is not allowed, i.e., perfectly complete allocations are enforced.

In [158], we discussed a way to detect and resolve such conflicts without relying on domain dependent information. This chapter is largely based on that work. Recall definitions 5.18 and 5.19, which describe a task allocation as a set of beliefs of the form $In(a, p, \tau, z)$. Task allocations are calculated for plantypes or hierarchies of plantypes by a task allocation algorithm. Afterwards, the corresponding beliefs are updated by received messages. As ALICA agents periodically broadcast their plan bases, locally computed task allocations are updated with parts of the result of remotely calculated ones. Alternatively, such an update can stem from action recognition, such that if the local agent is able to observe the actions of another, it can reason about its state within the plan-tree [75]. We treat such recognition tasks as communication by means of the environment.

6.1 Conflict Detection

As conflicts occur with respect to a plan or a plantype, in the following we will refer to allocations *for* a specific plantype P, which are subsets of belief bases limited to beliefs of the form $In(a, p, \tau, z)$, where $p \in P$. Furthermore, since states are not important for the following conflict detection scheme, we introduce the macro $In(a, p, \tau) \overset{def}{=} (\exists z) In(a, p, \tau, z)$.

Definition 6.1 (Conflict). Two allocations C_1 and C_2 are in conflict, if and only if they are for the same plantype P, and, for some agent a, it is not the case that $C_1 \vdash In(a, p, \tau) \leftrightarrow C_2 \vdash In(a, p, \tau)$. Two agents a_1 and a_2 are in conflict with respect to plantype P if and only if they believe in conflicting allocations. We say $In(a, p, \tau)$ is a *cause* of the conflict. Note that there can be multiple causes for a conflict.

Proposition 14. *If two agents a_1 and a_2 are in conflict with respect to plantype P, caused by* $\text{In}(b, p, \tau)$, *then there is a conflict between one of them and agent b.*

Proof. Suppose two agents a_1 and a_2 are in conflict about plantype P. Suppose furthermore, w.l.o.g., agent a_1 believes in a task allocation for P, which contains $\text{In}(b, p, \tau, z)$ for some state z and this is the cause for the conflict. Then it holds for the task allocation C_2 for P, a_2 believes in, that

- b is not allocated at all – $(\forall p', \tau', z') \, \text{In}(b, p', \tau', z') \notin C_2$, or

- b is assigned to another task τ' in $p - (\exists \tau', z') \, \text{In}(b, \tau', p, z') \in C_2 \wedge \tau \neq \tau'$, or

- b is assigned to another plan in $P - (\exists p', \tau', z') \, \text{In}(b, p', \tau', z') \wedge p \neq p'$

Since b cannot be allocated to two tasks within the same plan or two plans within the same plantype due to plan base consistency (Theorem 5.2), it cannot support the belief $\text{In}(b, p, \tau, z)$ without being in conflict with a_2. Since not supporting $\text{In}(b, p, \tau, z)$ causes a conflict with a_1, the claim holds. If b is either a_1 or a_2 the claim holds as well. □

This property guarantees that the agent, whose assignment is in dispute among the team is in conflict with at least one agent and thus has a chance to detect it.

In order to devise a detection mechanism, we examine how allocations are modified over time. An agent's allocation for a plantype P can change due to one of the following events:

- Reallocation – the agent adopts a new allocation to improve the utility.

- Message – the agent receives a message informing it about the state of another with respect to P.

- Leaving – the agent leaves the plan and thus no longer tracks its allocation.

- Deletion – the agent has not received a message from another agent for some time. In this case the agent is removed from the team and thus from all task allocations.

Deletion is done to account for agents breaking down and losing their ability to communicate. We will not consider the deletion event further, as it can be seen as a special case of the message event, namely as an empty plan base message. We will also not consider agents leaving a plan, this event terminates the local agent's tracking of the plan and is visible to other agents by message events. These are able to detect a conflict arising by an agent entering and leaving a plan repeatedly.

Persistent conflicts, i.e., those which endure for a longer period of time, cause a temporal pattern to appear in allocations believed by the involved agents. We therefore define formally how allocations change over time.

Definition 6.2 (Allocation Event). An allocation event e is a tuple $(\vartheta^+, \vartheta^-)$, consisting of allocation additions ϑ^+ and allocation subtractions ϑ^-, both sets of ground atoms of the form $\text{In}(a, p, \tau)$, such that $\vartheta^+ \cap \vartheta^- = \emptyset$ and $(\exists P)(\forall p)\,\text{In}(a, p, \tau) \in \vartheta^+ \cup \vartheta^- \rightarrow p \in P$, i.e., an allocation event contains only additions and subtractions with respect to a single plantype P.

We use the binary functor $\circ$ to denote the composition of allocation events, i.e., $e_1 \circ e_2 = e_3$.

Definition 6.3 (Allocation Event Composition). Let $e_1 = (\vartheta_1^+, \vartheta_1^-)$ and $e_2 = (\vartheta_2^+, \vartheta_2^-)$ be two task allocation events. Then their composition is defined as:

$$e_1 \circ e_2 \overset{def}{=} \left((\vartheta_1^+ - \vartheta_2^-) \cup (\vartheta_2^+ - \vartheta_1^-), (\vartheta_1^- - \vartheta_2^+) \cup (\vartheta_2^- - \vartheta_1^+)\right)$$

This composition allows reasoning about results of multiple events independently of the allocations they apply to. Allocation events form an Abelian group under event composition.

If an agent a receives a plan message from another agent b, this causes an allocation event for every plantype agent a is executing. There are four cases to distinguish:

- a agrees with b on its allocation, hence the allocation event is empty and can be ignored.

- a does not believe that b participates in the plantype, but b does, this yields the allocation event $(\{\text{In}(b, p, \tau)\}, \emptyset)$ for some p and τ.

- a believes that b executes a different plan or task than b actually does, yielding the event $(\{\text{In}(b, p, \tau)\}, \{\text{In}(b, p'\tau')\})$ such that $p \neq p' \vee \tau \neq \tau'$.

- a wrongly believes that b participates in the plantype, causing $(\emptyset, \{\text{In}(b, p, \tau)\})$ for some p, τ.

By Proposition 14, these are the only cases we need to consider. If a now reallocates and the composition of both events contains no statement about b, a cycle occurred.

Definition 6.4 (Allocation Cycle). Let e_m be a non-empty allocation event due to a plan message sent by agent b to agent a, and let e_r be the next allocation event after a receives the message. If $e_m \circ e_r = (A, B)$ such that $(\forall p, \tau) \operatorname{In}(b, p, \tau) \notin (A \cup B)$, a cycle occurred.

Hence, a cycle occurs, if an agent reverts its allocation with respect to a message by reallocation. An agent can easily monitor its allocation events and thus detect such cycles. Of course, a cycle does not necessarily entail a conflict, it might well be that the cycle just reflects a very quickly changing situation. However, if multiple cycles occur subsequently, this becomes more and more unlikely to be a proper reaction to the changing conditions. The cycle length depends on the frequency with which the agents communicate and deliberate. In robotic soccer, suitable values could be 10 Hz for the average communication frequency and 30 Hz for the deliberation loop. Thus, the duration of two subsequent cycles is in average 117 ms. It is improbable that during this time frame the situation requires the team to change its allocation back and forth twice. Thus, a limit $n \geq 2$ on the number of subsequent cycles can be established, above which an agent can assume with reasonable confidence that a conflict occurred. A reasonable limit can be found whenever both communication frequency and deliberation frequency are relatively high compared to the dynamic of the environment, and sent messages are received with a probability other than 0.

In the following we present an integration of cycle detection into the operational semantics of pALICA. We write $\Delta(P, B_1, B_2)$ to indicate the allocation difference between two belief bases B_1 and B_2 for a plantype P:

$$\Delta(P, B_1, B_2) \stackrel{def}{=} (\{\operatorname{In}(a, p, \tau) \mid p \in P \wedge (\exists z) \operatorname{In}(a, p, \tau, z) \notin B_1 \wedge \operatorname{In}(a, p, \tau, z) \in B_2\},$$
$$\{\operatorname{In}(a, p, \tau) \mid p \in P \wedge (\exists z) \operatorname{In}(a, p, \tau, z) \notin B_2 \wedge \operatorname{In}(a, p, \tau, z) \in B_1\})$$

In order to detect cycles, agents must remember parts of the allocation history. We assume the presence of two predicates in the belief base: $\text{Cycles}(P, i)$, indicating that i subsequent cycles occurred during the execution of plantype P, and $\text{ADiff}(P, d)$, which indicates that d is the currently remembered allocation difference for plantype P.

The necessary updates are formulated in the form of operational rules, extending the original operational semantics defined in Section 5.13.

We write

$$t : \frac{\operatorname{Conf}(a) \stackrel{r}{\to} \operatorname{Conf}(a)'}{(B, \Upsilon, E, R) \longrightarrow (B', \Upsilon', E', R')}$$

to indicate that if the rule r maps the configuration $\operatorname{Conf}(a)$ to $\operatorname{Conf}(a)'$, the rule t will map (B, Υ, E, R) to (B', Υ', E', R').

Firstly, whenever an agent learns new information regarding the state of other agents due to message reception or action recognition, the rule *Sense* incorporates that information into the belief base. The rule $Sense_{cd}$ extends this update to update history information represented by ADiff as well.

$$Sense_{cd} : \frac{(B,\Upsilon,E,R) \xrightarrow{Sense} (B',\Upsilon',E',R')}{(B,\Upsilon,E,R) \longrightarrow ((B'-\vartheta^-)+\vartheta^+,\Upsilon',E',R')}$$

where

- $\vartheta^- = \{\text{ADiff}(P,d) \mid (p,\tau,z) \in \Upsilon \wedge p \in P\}$

- $\vartheta^+ = \{\text{ADiff}(P,d \circ \Delta(P,B,B')) \mid \text{ADiff}(P,d) \in \vartheta^-\}$

Secondly, whenever an agent applies the rule *Adapt* to dynamically reallocate agents within a plan, the allocation difference is updated as well. Furthermore, should a non-empty reallocation yield an empty difference, a cycle is detected. This is captured by the rule $Adapt_{cd}$.

$$Adapt_{cd} : \frac{(B,\Upsilon,E,R) \xrightarrow{Adapt} (B',\Upsilon',E',R')}{(B,\Upsilon,E,R) \longrightarrow ((B'-\vartheta^-)+\vartheta^+,\Upsilon',E',R')}$$

where

- $A^- = \{\text{ADiff}(P,d) \mid (p,\tau,z) \in \Upsilon \wedge p \in P\}$

- $A^+ = \{\text{ADiff}(P,d \circ \Delta(P,B,B')) \mid \text{ADiff}(P,d) \in A^-\}$

- $B^- = \{\text{Cycles}(P,i) \mid \text{Cycles}(P,i) \in B \wedge \text{ADiff}(P,(\gamma^+,\gamma^-)) \in A^- \wedge$
  $\qquad\qquad (\gamma^+ \neq \emptyset \vee \gamma^- \neq \emptyset) \wedge \text{ADiff}(P,(\emptyset,\emptyset)) \in A^+\}$

- $B^+ = \{\text{Cycles}(P,j) \mid \text{Cycles}(P,i) \in B^- \wedge j = i+1\}$

- $C^- = \{\text{Cycles}(P,i) \mid \text{Cycles}(P,i) \in B \wedge \text{Cycles}(P,i) \notin B^- \wedge$
  $\qquad\qquad \Delta(P,B,B') \neq (\emptyset,\emptyset)\}$

- $C^+ = \{\text{Cycles}(P,0) \mid \text{Cycles}(P,i) \in C^-\}$

- $\vartheta^+ = A^+ \cup B^+ \cup C^+$

- $\vartheta^- = A^- \cup B^- \cup C^-$

That is, whenever a reallocation is performed, the corresponding allocation differences are updated. Should an update yield an empty allocation difference, the corresponding cycle count is increased. If a reallocation yields a non-empty difference, the cycle count is reset to 0, as in this case the sequence of cycles ended, and the adaptation is probably due to the dynamics of the environment.

Finally, whenever an agent leaves a plantype, information about allocation differences and cycles are reset. Thus, the next rule, $Reset_{cd}$, resets the corresponding information, whenever a rule r from the original set of rules, denoted by $pALICA$ stops the execution of the plantype.

$$Reset_{cd} : \frac{(B,\Upsilon,E,R) \xrightarrow{r \in pALICA} (B',\Upsilon',E',R')}{(B,\Upsilon,E,R) \longrightarrow ((B' - \vartheta^-) + \vartheta^+, \Upsilon', E', R')}$$

where

- $\vartheta^- = \{Cycles(P,i), ADiff(P,d) \mid P \in \mathcal{P}_\vee \wedge \neg(\exists p)(p,\tau,z) \in \Upsilon' \wedge p \in P\}$

- $\vartheta^+ = \{Cycles(P,0), ADiff(P,(\emptyset,\emptyset)) \mid P \in \mathcal{P}_\vee \wedge \neg(\exists p)(p,\tau,z) \in \Upsilon' \wedge p \in P\}$

6.2 Conflict Resolution

Given the detection scheme discussed in the previous section, we can devise a reaction to observed conflicts. Once an agent has detected n consecutive cycles, it can assume the presence of a conflict. In order to overcome a conflict, we switch the task allocation procedure for the plantype in conflict to a central coordinator scheme. Such a local leader concept is similar to the team leaders used in STEAM [163], however, here we only revert to a leader-based decision making approach in case of conflicts, not as a default handling mechanism.

One of the simplest method to elect a leader is bullying [54]. Bullying requires nodes to broadcast election messages if they deem themselves a leader. A total order over the potential participants dictates which out of the active nodes becomes leader. Therefore, whenever a node receives a message from a node with lower precedence, it starts broadcasting, should it receive a message from a node with higher precedence, it stops. Note that in case of unreliable, asynchronous communication, a node can never be sure to be the only one that considers itself a leader. However, we can make the assumptions that messages are either correctly received or not received at all, i.e., messages are not distorted by the communication channel, and that not all messages are lost. We deem these assumptions reasonable, as one can use appropriate error detection codes to recognise and discard distorted messages. Based on these assumptions, a continuous bullying algorithm, in which a node that deems itself leader never stops broadcasting, can guarantee that a single node will eventually become the unique leader of the group in question.

In our case, the leader periodically broadcasts an allocation for a plantype. We can therefore combine the election message with the actual content and thereby reduce communication overhead. In the remainder, we refer to these combined messages as *authority messages*. The resulting protocol allows an agent to immediately react to a received authority message. Upon reception of such a message, an agent has three options: Firstly, it starts to broadcast authority messages of its own, if it has not received an authority message from an agent of higher precedence with respect to the bullying order. Secondly, it adapts the contained allocation in case the authority message has the highest precedence out of the messages it received so far. Thirdly, it ignores the message, if it knows of a leader with higher precedence.

When the team's allocation is determined by a leader, it is less reactive than in the normal, distributed decision-making state. This is due to the additional messages which need to be sent by the leader and the fact that the leader has to be informed of any situation change which it cannot observe itself. Therefore, the team should revert back to its more dynamic mode as soon as possible. On the other hand, if the team switches back too quickly, the source of the conflict, such as an inconsistency in the belief bases, may still persist, and the conflict will reoccur. Therefore, the time interval in which the team is in authoritative state with respect to a plantype should be adapted dynamically.

A corresponding algorithm is depicted in Listing 6.1. Each agent maintains for each plantype the mode of coordination, Mode(P), and a varying time interval, AuthTime(P), indicating the time the team spends in authoritative mode. This time varies between $t_{\min}$ and $t_{\max}$, depending on how frequently conflicts are detected. The precedence among agents can be chosen arbitrarily, but must be common knowledge within the team. Unique identifiers from a totally ordered set are a typical choice for bully algorithms.

Incorporating this algorithm into the operational semantics is straight forward. Firstly, additional beliefs are needed to reflect the state of each plantype with respect to conflict resolution. Thus we assert the following predicates into the belief base:

- *Mode*(P,x) – indicating that plantype P is currently executed in mode x.

- *AuthTime*(P,t) – indicating that the current authority time for plantype P is t.

```
foreach ( plantype  P  in  execution ) do {
  if  (Mode(P) = normal) {
      if  ( conflict  detected  or  incoming authority  message with lower precedence) {
          Mode(P) :=  authority ;
          t_begin := now;
          AuthTime(P) := min(t_max,AuthTime(P)·t⁺);
      }
      if  (incoming authority  message with  higher  precedence)
          Mode(P) := commanded;
      else {
          perform  reallocation  if  needed;
          AuthTime(P) := max(t_min,AuthTime(P)·t⁻);
      }
  }
  else  if  (Mode(P) = authority ) {
      perform  reallocation  if  needed;
      broadcast  authority  message;
      if  (AuthTime(P) < t_begin) Mode(P)) := normal;
      if  (incoming authority  message with  higher precedence) Mode(P) := commanded;
  }
  else {
      adopt  allocation  of  last  received  message with  highest precedence;
      if  (time  since  last  authority  message > t_timeout)
          Mode(P) := normal;
  }
}
```

Listing 6.1: Conflict Resolution Algorithm

Additionally, the following axioms are included in Σ_b:

$$(\forall P \in \text{PlanTypes}^+(z_0))(\exists x)Mode(P,x)$$
$$(\forall P \in \text{PlanTypes}^+(z_0))(\exists t)AuthTime(P,t) \wedge t > 0$$
$$Mode(P,x) \wedge Mode(P,y) \rightarrow x = y$$
$$AuthTime(P,t) \wedge AuthTime(P,s) \rightarrow t = s$$
$$Mode(P,x) \rightarrow x = normal \vee x = commanded \vee x = authority$$

The first two predicates are functional in their second argument, and for each plantype there is exactly one positive occurrence of this predicate in the belief base. Furthermore, the only modes allowed for execution are normal, commanded, and authority, expressing that the respective plantype is executed normally, under the authority of another agent, or as a leader, respectively.

In the following we extend the set of rules discussed so far to include conflict resolution. Most importantly, the rule *Adapt* is modified such that it only applies to plantypes not currently executed in commanded mode. That is, dynamic reallocation is only done in normal mode or by the leader in authoritative mode.

$$Adapt : \frac{Mode(P,x) \wedge x \neq commanded \wedge (B,\Upsilon,E,\mathrm{R}) \xrightarrow{Adapt} (B',\Upsilon',E',\mathrm{R}')}{(B,\Upsilon,E,\mathrm{R}) \longrightarrow ((B' - AuthTime(P,t)) + AuthTime(P,t_n), \Upsilon',E',\mathrm{R}')}$$

where P is the plantype in which a reallocation is triggered and $t_n = \max(t_{\min}, t \cdot t^-)$.

An agent assumes authority, whenever the number of subsequent cycles detected reaches the set threshold n. In this case, the mode switches to authority, the authority time is increased, and the number of cycles is reset. The rule $Auth_{cr}$ captures this behaviour.

$$Auth_{cr} : \frac{(p,\tau,z) \in \Upsilon \wedge p \in P \wedge Mode(P,normal) \wedge AuthTime(P,t) \wedge \phi}{(B,\Upsilon,E,\mathrm{R}) \longrightarrow ((B - \vartheta^-) + \vartheta^+, \Upsilon,E,\mathrm{R})}$$

where

- $\phi = \mathrm{Cycles}(P,x) \wedge x \geq n$

- n is a fixed threshold, indicating that n subsequent cycles according to Definition 6.4 occurred.

- $\vartheta^- = \{Mode(P,normal), \mathrm{Cycles}(P,x), AuthTime(P,t)\}$

- $\vartheta^+ = \{Mode(P,authority), \mathrm{Cycles}(P,0), AuthTime(P,\min(t_{\max}, t \cdot t^+))\}$

The core of conflict resolution by authority is the rule Cmd_{cr}, which reacts to incoming authority messages. The event of such a message reception is denoted by $AuthMsg(a',P,C)$, where a' is the sending agent, P the plantype in question, and C the task allocation a' proposes. If the receiving agent considers the sending agent of higher precedence ($a < a'$), and participates in P, it will assume the sent allocation, and if it thereby enters a new state it will schedule a task allocation for the newly entered state.

$$Cmd_{cr} : \frac{AuthMsg(a',P,C) \wedge a < a' \wedge \mathrm{In}(a,p,\tau,z) \wedge p \in P}{(B,\Upsilon,E,\mathrm{R}) \longrightarrow ((B - \vartheta_b^-) + \vartheta_b^+ + \vartheta_b', (\Upsilon - \vartheta_p^-) \cup \vartheta_p^+, (E - \vartheta_e^-), \mathrm{R})}$$

where

- a indicates the local agent

- $\vartheta_b^- = \{Mode(P,x), Cycles(P,c)\} \cup \{In(a'', p', \tau', z') \mid p' \in P\}$

- $\vartheta_b^+ = \{Mode(P, commanded), Cycles(P,0)\} \cup C$

- $\vartheta_b' = \begin{cases} \{Alloc(z')\} & \text{if } In(a, p', \tau', z') \in C \wedge z \neq z' \\ \emptyset & \text{otherwise} \end{cases}$

- $\vartheta_p^- = \begin{cases} \{(p, \tau, z))\} \cup \{(p', \tau'', z'') \mid p' \in \text{Plans}^+(z)\} & \text{if } Alloc(z') \in \vartheta_b^+ \\ \emptyset & \text{otherwise} \end{cases}$

- $\vartheta_p^+ = \{(p', \tau', z') \mid In(a, p', \tau', z') \in C\}$

- $\vartheta_e^- = \{(b, z_d) \mid (p_d, \tau_d, z_d) \in \vartheta_p^- \wedge b \in \text{Behaviours}(z_d)\}$

If, on the other hand, the receiving agent considers itself of higher precedence, it will in turn switch to authority mode and start to broadcast task allocations, according to the bully protocol:

$$TakeAuth_{cr} : \frac{AuthMsg(a', P, C) \wedge a > a' \wedge In(a, p, \tau, z) \wedge p \in P \wedge Mode(P, normal)}{(B, \Upsilon, E, R) \longrightarrow ((B - \vartheta_b^-) + Mode(P, authority), \Upsilon, E, R)}$$

where $\vartheta_b^- = Mode(P, normal)$.

When the time threshold for authority is reached, the leader will drop back to normal mode, and stop broadcasting. Although not required, it can notify the team of this mode switch by an additional message, which can speed up mode switch for the rest of the team. Otherwise, commanded agents will register the lack of authority messages and switch back after a timeout has been reached, due to rule $DropCmd_{cr}$. This timeout is essential, since the leading agent can break down at any time, leaving the team without a leader.

$$DropAuth_{cr} : \frac{Mode(P, authority) \wedge AuthTime(P, t) \wedge \phi(P, t)}{(B, \Upsilon, E, R) \longrightarrow ((B - \vartheta_b^-) + Mode(P, normal), \Upsilon, E, R)}$$

where $\vartheta_b^- = Mode(P, authority)$ and $\phi(P, t)$ holds if the time since the mode for P was switched to authority is larger than t.

$$DropCmd_{cr} : \frac{Mode(P, commanded) \wedge \phi'(P, t_{timeout})}{(B, \Upsilon, E, R) \longrightarrow ((B - \vartheta_b^-) + Mode(P, normal), \Upsilon, E, R)}$$

where $\vartheta_b^- = Mode(P, commanded)$ and $\phi'(P, t_{timeout})$ holds if the time since the last authority message received is larger than $t_{timeout}$. The constant $t_{timeout}$ is a parameter that depends on the communication frequency, and the expected packet loss.

Finally, the information represented by $Mode(P, x)$ needs to be reset whenever the local agent leaves a plantype. This is modelled by the rule $Reset_{cr}$.

$$Reset_{cr} : \frac{(B, \Upsilon, E, R) \xrightarrow{r \in pALICA} (B', \Upsilon', E', R')}{(B, \Upsilon, E, R) \longrightarrow ((B' - \vartheta^-) + \vartheta^+, \Upsilon', E', R')}$$

where

- $\vartheta^- = \{Mode(P, x) \mid (p, \tau, z) \in \Upsilon \wedge p \in P \wedge \neg(p' \in P \wedge (p', \tau', z') \in \Upsilon')\}$

- $\vartheta^+ = \{Mode(P, normal) \mid Mode(P, x) \in \vartheta^-\}$

This set of rules enables a team to react to conflicts dynamically and adaptive by creating a leader responsible for the task allocation for a single node, i.e., plantype within the plan-tree. Most notably, the leader is only elected with respect to a specific node in the plan-tree, thus operation of the rest of the tree is unaffected by conflict resolution and continues in the normal dynamic way. Of course, other schemes of agreement are possible, for instance using a majority vote. Such schemes can be integrated into ALICA in a similar fashion.

It is easy to see that this conflict detection and resolution scheme maintains the properties discussed in Section 5.14.

Proposition 15. *The rule extension maintains consistency with the plan base axioms Σ_p and maintains the tree shape of the plan base.*

Proof. The only new rule that modifies the plan base is Cmd_{cr}. It adapts a valid allocation, since only valid allocations are sent by the leader. In cases where the leader cannot find a valid allocation, it would abort the corresponding plan. Cmd_{cr} removes the current sub-branch from the plan base in case the local agent is reallocated by the leader, in which case it adds at most one tuple to the plan base and triggers a recursive allocation in the newly entered state. This tuple refers to a plan in the plantype P. Since any tuple referring to a plan in the plantype P was removed, the claim holds. □

Proposition 16. *The rule extension does not leave any orphaned behaviours.*

Proof. None of the newly introduced rules adds a behaviour to the execution set. The only rule that modifies the plan base is Cmd_{cr}, which removes all tuples (b, z)

from the execution set if it removes the corresponding triple (p, τ, z) from the plan base.   □

Proposition 17. *The rule extension maintains Proposition 10.*

Proof. The only rule that modifies the plan base or beliefs of the form $In(a, p, \tau, z)$ is Cmd_{cr}, which updates both in the same way.   □

We will present an empirical evaluation of this conflict resolution technique in Section 10.2. Note that it is not possible to solve all conflicts in this way. Conflicts resulting from unsound plans or not perfectly sound plans executed without allowing for agents to idle cannot be resolved in this way. A resolution to such issues by authority requires the leading agent to consider sub-plans it does not take part in. This could be achieved by switching to a global task allocation whenever a leader is elected. In this case, the leader would calculate the recursive task allocation for all participating agents with the conflicting plantype as root node. However, as mentioned earlier in Section 5.9, the leader might not have sufficient information to do that. Alternatively, the leader is informed by team members of conflicts resulting from requirements in lower-level plans. Agents could broadcast information indicating that they find themselves unable to perform the task required, whenever it cannot comply with an authority message. Subsequently, the leader can take that information into account. However, since potentially multiple of such messages need to be exchanged in order to find a suitable allocation, such an approach might result in an outdated allocation due to the dynamic environment. We consider investigating such techniques as future work. Ideally, this kind of conflict is avoided entirely, since conflict resolution always degrades performance to a certain degree. Instead, the soundness of plans should be verified in a pre-runtime validation step or be guaranteed by the generating process, e.g., by the employed planning algorithm. Table 6.1 summarises the beliefs and rules introduced in this chapter in order to express conflict detection using allocation cycles and conflict resolution by means of a local leader.

Beliefs	
ADiff(P,d)	ADiff(P,d) identifies for each plantype P the current allocation difference. If the difference becomes empty due to an application of the rule *Adapt*, a cycle occurred.
Cycles(P,i)	Cycles(P,i) denotes how many subsequent allocation cycles were detected while executing of plantype P.
Mode(P,x)	x is the mode in which plantype P is executed. The mode can be either normal, authority, or commanded.
AuthTime(P,t)	This predicate reflects for each plantype P the dynamically adaptable duration of a conflict resolution.
Rules	
Sense$_{cd}$	This rule modifies the rule *Sense* such that allocation differences are recorded in the belief base.
Adapt$_{cd}$	Adapt$_{cd}$ modifies the rule *Adapt* to update allocation differences and cycle count.
Reset$_{cd}$	Reset$_{cd}$ resets information kept about plantypes no longer in execution.
Adapt	The original rule *Adapt* is modified such that commanded agents do not modify their allocation autonomously.
Auth$_{cr}$	Auth$_{cr}$ switches an agent to authority mode whenever it detects a sufficient amount of subsequent cycles.
Cmd$_{cr}$	Cmd$_{cr}$ enforces authority by switching the local agent to commanded mode whenever an authoritative message from an agent with higher precedence arrives. In this case, the contained allocation is enforced.
TakeAuth$_{cr}$	TakeAuth$_{cr}$ implements the bully rule, that is whenever an authoritative message from an agent with lower precedence is received, the agent switches to authority mode.
DropAuth$_{cr}$	DropAuth$_{cr}$ forces a leading agent to drop back to normal mode once the duration indicated by *AuthTime* is reached.
DropCmd$_{cr}$	DropCmd$_{cr}$ expresses that an agent returns to normal operation from commanded operation after no authority message was received for a certain time.
Reset$_{cr}$	Reset$_{cr}$ resets the mode of a plantype the agent stops executing, regardless of the reason.

Table 6.1: Beliefs and Rules Used to Express Conflict Detection and Resolution

7 Software Architecture

Although software architecture considerations are not at the heart of this work, the complexity of multi-agent coordination and the diversity of possible approaches warrants a discussion. The ALICA reference implementation serves as a basis for this discussion. This reference implementation is publicly available[1] under a BSD-based license.

An ALICA engine is a software module, that executes an ALICA program. It maintains a consistent state of the plan base, as well as all beliefs related to it and discussed here, triggers behaviours, and communicates with other agents running ALICA engines. In order to do that, the ALICA program needs to be available in some machine readable form. This form is provided by a modelling tool chain, which we will discuss in Section 7.1. Afterwards, we discuss the internal layout of the reference ALICA engine in Section 7.2. In Section 7.3, we discuss a possible model for the overall architecture, in which ALICA can be embedded.

7.1 Modelling Tools and Exchange Format

The ALICA reference implementation features a graphical development environment, called the *ALICA PlanDesigner*, which allows the creation of all ALICA language elements. The original implementation was developed by Scharf [144] in 2008 and was since then continuously advanced.

The practical representation of ALICA is an XML-based language, which can be used as interchange format between modelling tools, execution engines, and model checkers. This representation is a domain-specific language (DSL), implemented with the Eclipse Modelling Framework (EMF) [161]. The EMF model, specified in UML [116], gives rise to the XMI [76] serialisation used as interchange format between editor and engine.

The PlanDesigner is a graphical tool based on the Eclipse Development Platform [7]. It supports modelling of all parts of an ALICA program, i.e., roles, tasks, plans, plantypes, utility functions, and conditions, as well as to generate code from the models in a model-driven development fashion. As described before, modelled plans are stored in the XMI format and loaded afterwards by the

---

[1]   http://ros.org/wiki/cn-alica-ros-pkg

runtime engine. However, for efficiency reasons, the tool provides mechanisms for generating platform-specific code for the evaluation of conditions and utilities. Since these evaluations happen very frequently during runtime, the generation of platform-specific code, which can be executed directly, results in enormous efficiency benefits. In order to facilitate an intuitive understanding, language elements are represented graphically.

7.2 Engine Layout

The ALICA reference engine is implemented in C#, and runs under mono.[1] C# is a strongly-typed, imperative, object-oriented language. C# programs are commonly compiled just-in-time[2] and use a garbage collector.[3] C# lends itself to reference implementations, since the imperative object-oriented programming paradigm is widely used and known [15]. Further, C# is easier to read and understand than C++ [174], while still retaining efficiency [56]. Finally, it contains additional features not present in older languages like Java, such as delegates, limited operator overloading, generics over value types, lambda expressions, and properties (see also [24]).

The reference engine is meant to be integrated into a domain-specific framework, which comprises of other components such as a belief representation framework, a communication middleware, sensor and actuator drivers. As such, it is built as a library providing an extensive API.

Figure 7.1 depicts a very general layout of the reference implementation. It comprises of five central components and a set of additional modules.

Engine Interface The Engine Interface provides all necessary functionality to start the engine, access different components, and configure its runtime behaviour.

PlanBase Central to the execution of an ALICA program is the PlanBase. It contains the runtime representation of the program, i.e., the current state of the agent within the plan structure. In its main loop, it takes care that all relevant components and modules have subsequent access to the plan structure.

---

[1]   www.mono-project.com
[2]   See for instance the work by Aycock [4] and Arnold et al. [3] for an overview on just-in-time compilation and other techniques.
[3]   Garbage Collectors are discussed in-depth in [79].

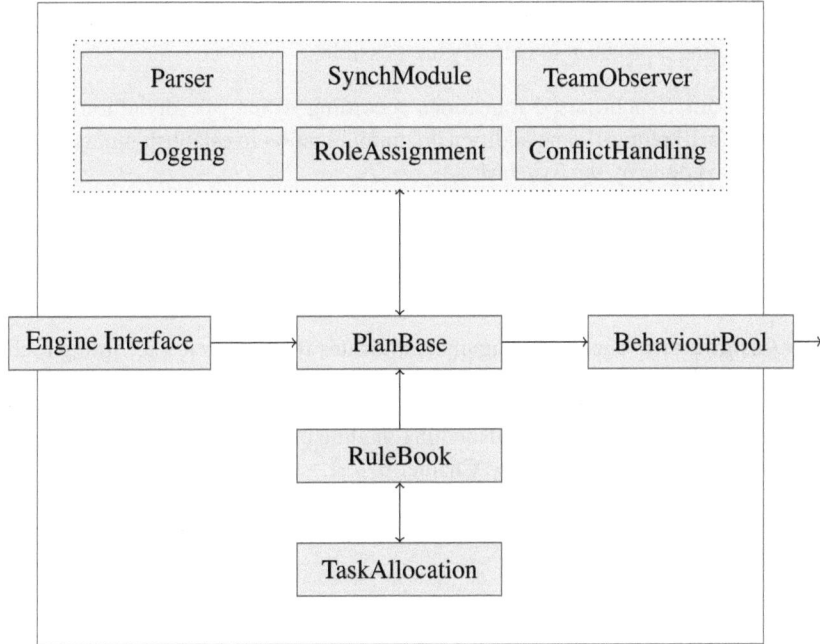

Figure 7.1: The ALICA Engine Reference Architecture

RuleBook The RuleBook contains the operational rules of ALICA. Called periodically by the PlanBase, it updates the plan-structure according to the rules.

TaskAllocation The TaskAllocation contains the algorithms discussed in Section 5.8. If needed, the RuleBook uses these task allocation algorithms to update the plan structure.

BehaviourPool This component manages the set of behaviours available to the agent. It calls behaviours currently in the execution set periodically or event driven, depending on the specification of the behaviour. It also supplies behaviours with all necessary parameters.

Apart from these essential components, the engine manages an extensible set of important, but replaceable modules.

Parser This module builds the internal representation of the ALICA program given the XML-documents and libraries built by the modelling tools discussed in Section 7.1. Although this an essential task, other means of constructing a program in memory are possible.

SynchModul Synchronised transitions according to the rule definition in Section 5.13 require some additional communication to establish mutual belief. This is done by the SynchModul.

TeamObserver The TeamObserver handles all periodic communication with team members, holds the internal representation of their current state within the program and makes estimations based on received messages.

RoleAssignment The RoleAssignment allocates roles to agents according to the allocation algorithm discussed in Section 5.6.

ConflictHandling The ConflictHandling module detects and resolves conflicts in task allocation according to Chapter 6.

Logging The logging module provides a simple facility to log rule applications to a file.

7.3 Agent Software Architecture

In this section, we propose agent architecture based on the agent model discussed in Section 5.2. The proposed architecture is largely based on deliberations by Baer [5] and the domain-specific architecture CANOSA described in his work and in [1].

Figure 7.2 shows the layout of the proposed architecture. Information stemming from sensors are processed by a single or potentially multiple external *Perception* components. These perform various tasks, such as image processing or sensor fusion. Abstract information is sent to the *WorldModel*. The WorldModel is a process-internal representation of the agent's belief state. Although in some scenarios it is unnecessary to maintain an internal representation of the world, in most real-world scenarios, the environment as well as the tasks that need to be performed in it become complex. Thus, an internal representation of the beliefs is more than helpful. We do not impose any restrictions as to how beliefs are represented and indeed, the ALICA engine does not access the WorldModel directly. Hence, any representation suitable to the domain can be used, be it BDI-based belief models according to Bratman [12], knowledge representation models such as employed by FLUX [168], ontology-based models (c.f. [25, 41]), or just plain data structures holding the most recent sensory information.

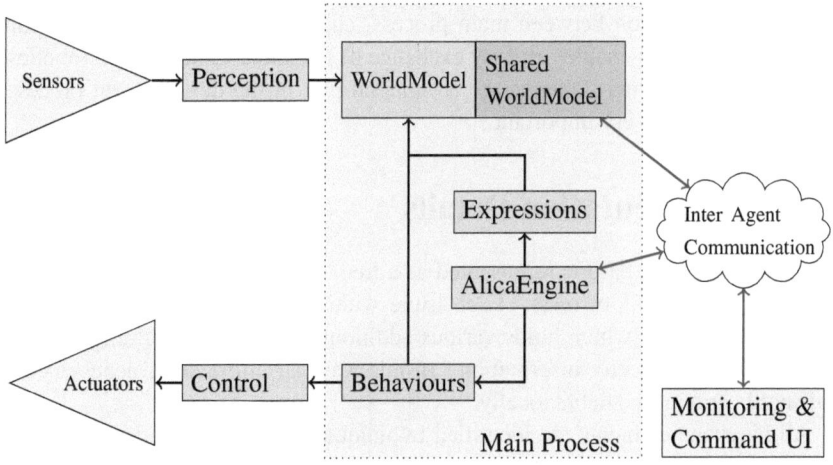

Figure 7.2: Agent Architecture

The WorldModel can be extended with a *SharedWorldModel*, which integrates information coming from other agents. The task of the SharedWorldModel is two-fold, firstly it estimates the knowledge and belief of other agents, secondly it fuses information received with local information in order to obtain a unified, consistent view on the world. Reichle [131] discussed this topic in detail.

As mentioned, the AlicaEngine does not access information in the WorldModel directly, but relies on two domain-specific components, namely the *Expressions* and the *Behaviours*. The Expressions component holds the implementations of all formulae and utility functions used in plans. As discussed in Section 7.1, these are not part of the XMI-based representation of the ALICA program, but are separate code fragments, since this potentially increases efficiency while maintaining a loose coupling between engine and belief representation. Beliefs regarding the internal state of agents, such as $In(a, p, \tau, z)$ and $Succeeded(a, p, \tau)$ are managed internally by the AlicaEngine, which provides suitable interfaces to access these beliefs.

The Behaviours component contains all domain-specific behaviours used in the ALICA program. These are controlled by the BehaviourPool within the engine. In turn, each behaviour communicates with actuation controllers, situated externally in the *Control* component. These controllers then communicate with the robotic hardware or software actuators.

This strict division between main process, and additional sensor or actuator-related components enables an easy exchange of the latter. Especially in robotics, where simulation environments are an integral part of the development process, exchangeability is very important.

7.4 Implementation Details

Internally, the plan base is represented as a tree, which is traversed by the Rule-Book in order to apply rules. Each triple within the plan base corresponds to a node in this tree, which holds various additional data structures, ranging from timestamps to authority information. In this way, all information necessary for rule application is available locally.

All language elements are identified by unique 64-bit identifiers and stored in hash maps, such that agents only need to communicate these identifiers in order to refer to plans, tasks, or states. Thus serialisation and deserialisation overhead are kept at a minimum.

Each behaviour is identified with a thread, that is constantly in stand-by until the behaviour is executed, reducing start-up time for individual behaviours significantly. All threads within the engine are able to run asynchronously to each other in regular intervals, or react on signals, which are exposed via the EngineInterface. Thereby, the engine's runtime behaviour can easily be adopted to accommodate for the specific needs of a domain. For example, a behaviour setting a success signal ($\mathrm{Success}(b,z)$) immediately wakes the PlanBase thread, which in turn applies all applicable rules in sequence, which can lead to another behaviour becoming active. This eliminates latency due to asynchronously running threads.

7.5 Communication

Each agent communicates with the team via some inter-agent communication method, commonly based on a middleware. The ALICA reference implementation is based on ROS (Robot Operating System) by Quigley et al. [127], which among many other features provides communication mechanisms based on a publish-subscribe model. However, since ROS uses a master process to establish and manage communication channels, it is currently by itself not usable in a multi-robot scenario, since this master becomes a single point of failure. In order to overcome this problem, proxies to other middleware systems such as the Data Distribution Service (DDS) [65] can be used. In the robotic soccer domain, we use a simple UDP multicast proxy due to efficiency reasons.

7.5.1 Information Exchange

Information exchanged between pALICA agents can be divided into two categories: domain-specific information regarding the environment and ALICA internal messages. The domain-specific messages typically reflect the sender's belief base or parts of it. This part of the communication is not governed by ALICA and is assumed to take place through some potentially unreliable communication channel.

pALICA defines the following communication means:

- A status broadcast, containing an agent's plan base, together with relevant success information,

- A handshake-protocol to establish mutual belief for synchronised transitions.

- Allocation authority messages, containing allocations for specific plans for conflict resolution according to Chapter 6

Although ALICA is meant to be a domain-independent language for multi-agent systems, it has a certain emphasis on mobile robotic domains, where communication is typically realised wireless. Thus, communication uses an unreliable broadcast medium, which lends itself well to broadcast messages as employed by ALICA. We assume the cost of a broadcast is equivalent to the cost of a peer-to-peer message.

The status broadcast is sent periodically with a dynamically chosen frequency of either f_{max} or f_{min}, depending on whether the internal state has changed recently or not. The status message contains the internal state of the agent, i.e., its plan base, and the task-plan tuples it successfully completed and still believes to be relevant according to Definition 5.6. Failures are not explicitly sent as these are apparent due to the agent no longer executing the corresponding plan. Success information however must be made explicit, in order to distinguish the two cases. If in agent is notified of a failure in this way, it can either try to compensate for the missing agent, or abort execution of the respective plan itself.

The size of these status messages depends on whether or not states are reachable from different initial states, i.e., whether or not the state machines contained in plans are connected or disjoint. In case they are disjoint, a state unambiguously identifies the corresponding task. This potentially halves the message size. The reference implementation therefore assumes disjoint state machines.

In practice, it is not always ideal to accept an incoming plan base message, due to message delay. For instance, after an agent performed a dynamic reallocation it can expect that messages arriving just afterwards will contradict it, as they were sent before the corresponding event in the environment was detect. Therefore, it is sensible to ignore such contradictions for a short, delay depending time interval after (re-)allocations.

The synchronisation protocol is implemented as a three-way handshake. Each participating agent announces its readiness to synchronise if it inhabits a state z such that a transition $t = (z, z', \phi)$ is an element of a synchronisation s and if it believes ϕ to hold. Each participating agent acknowledges the announcement, and broadcasts a readiness signal once it received an acknowledgement of all announcements from every participating agents. In that moment it starts to support the mutual belief and is able to follow the synchronised transition according to the rule *STrans*. Should an agent receive a readiness signal while believing its respective condition ϕ holds, it immediately supports the mutual belief. Should contradictory information arise during the establishment of the belief, the corresponding agent informs the group to retract its commitment.

Establishing mutual belief in a synchronised manner through communication comes down to solving the coordinated attack problem [62, 58], which is known to have no finite solution, given asynchronous, unreliable communication. Therefore, any protocol for synchronised transitions can only be approximately correct. We deem a three-way handshake acceptable under most conditions, however, it is easy to extend this basic protocol to an n-way handshake.

Allocation authority messages are broadcast periodically by an agent which deems itself the leader with respect to a certain plantype according to the detection and resolution scheme discussed in Chapter 6. Each message identifies the sender, such that bullying can take place. Furthermore, each message contains the full allocation for the plantype in question. Should the leader stop to send its messages, the conflict is deemed to have been resolved and the team drops back its normal mode of operation. Thereby, the potential break down of a leader is handled as well.

7.5.2 Estimating the Current Team

For a team to function in a coherent manner, it is of crucial importance to estimate the set of agents participating in the team, i.e., the current team composition. While ALICA assumes that the whole team of agents, $\mathcal{A}$, is known in advance, this does not mean that all agents potentially participating are actually available. Moreover, agents may be incapacitated anytime. Thus, each agent needs to keep track of the *active team*, which is defined to be the set of agents participating in the top-level plan, $\{a \mid B \vdash \text{In}(a, p_0, \tau_0, z_0)\}$.

When agent a receives a message from agent b, it can conclude that agent b was active when it sent the message. Assuming all robots are within communication range to each other, it can also make estimations based on the number of expected, but not received messages. Since ALICA agents periodically broadcast messages with a frequency of f_{min} or f_{max}, the number of missing messages ranges between $t \cdot f_{min}$ and $t \cdot f_{max}$ where t is the time since the last message reception, assuming communication jitter is negligible. If the difference between f_{min} and f_{max} is relatively small, an agent can estimate the number of expected messages as $\tilde{n} = 0.5t(f_{min} + f_{max})$.

The more messages from a specific agent are expected, but not received, the lower the probability that this is due to messages being lost. Given the probability of a message sent being delivered, p_m, the probability that the missing messages are due to communication errors can be roughly estimated by

$$p_{loss} = (1 - p_m)^{\tilde{n}}$$

However, this is a very crude approximation. Network errors typically appear in bursts, thus the probabilities for each message are not independent of each other. Messages sent consecutively by an agent are likely subject to the same error burst. Moreover, messages sent by different agents at the same time are also likely subject to the error.

There has been extensive research on the modelling of burst errors and their probability distribution. A widely used description is the Neyman-A contagious model [109], which was originally used to describe the distribution of larvae in experimental field tests. Using this model, the length of burst errors can be described with a Poisson-distribution (cf. [44]). Thus, we can model the probability of the loss of $\tilde{n}$ subsequent messages as the probability of a burst error of length $k \geq \tilde{n}$:

$$p(k \geq \tilde{n}) = 1 - e^{-\lambda} \sum_{i=0}^{\tilde{n}-1} \frac{\lambda^i}{i!} \tag{7.1}$$

Assuming an a priori probability estimation for a robot breaking down, $p_{down}(a)$, is available, each agent can estimate the active team by:

$$\{a \mid B \vdash \text{In}(a, p_0, \tau_0, z_0)\} = \{a \mid p_{down}(a) < p(k \geq \tilde{n}(a))\} \tag{7.2}$$

where $\tilde{n}(a)$ is the number of consecutive messages expected but not received from agent a. Equation 7.2, together with Σ_B dictates that an agent removes all believes of the form $\text{In}(a, p, \tau, z)$ from its belief base if it estimates $p_{down}(a) \geq p(k \geq \tilde{n}(a))$. Thus, if for example $p_{down}(a) = 10^{-3}$, $f_{max} = 15$, $f_{min} = 5$, and $\lambda = 10$ an agent will assume that another agent a broke down after 22 consecutive messages missing, i.e., after 2.2 s of not receiving any message from a. In the robotic soccer domain for instance, these values are a reasonable assumption.

Note that this estimation does not model any dependencies between messages received from different agents. Moreover, when agents are expected to move in and out of communication range for extended periods of time, this simple approximation will no longer suffice. Furthermore, there may be additional evidence available besides the reception of messages or lack thereof. An agent might be able to sense another agent's effect on the environment and use that information to draw conclusions about its internal state. See for instance Huber and Durfee [75] for a discussion on how joint intentions can be achieved without communication. Here, we assume all information perceived regarding the state of another agent can be expressed in terms of communication, be it explicit (through for instance a wireless communication link), or implicit through action effects on the environment.

In mobile robotic scenarios, where robots can move out of communication range, the probability of receiving a message is a function of their positions in the environment.

$$p_m(\Delta t) = \int_{t_0}^{t_0 + \Delta t} f(pos(a,t), pos(b,t)) dt$$

where t_0 denotes the time of the last message reception, $pos(a,t)$ and $pos(b,t)$ the positions of the sender and receiver at time t respectively and $p_m(\Delta t)$ a probability distribution of a message being successfully delivered during Δt given the positions of sender and receiver within the environment. The probability distribution can be arbitrarily complex if the environment features objects which reflect radio signals or sources of interference. Accurate approximations of such probability distributions can be obtained using ray tracing and other simulation techniques but are computationally expensive to obtain in complex environments (see for instance [114, 145]). Such elaborate models are usually not suitable for real-time capable probability estimations. Instead, faster methods are sought after, which sacrifice accuracy for speed. These usually only take line-of-sight and distance into account, estimate the probability based on the signal-to-noise ratio [148], or use test packages to measure the link quality [87]. For the purpose of this work, we limit ourself to the simple estimation in Equation 7.1.

7.6 Summary

In this chapter, we described the ALICA engine architecture, which follows the theoretical foundations described in the previous chapter. Moreover, a modelling tool and an exchange format for ALICA programs were described. While the AL-ICA engine can be embedded in any agent architecture, we discussed necessary features on the basis of a general architecture. Finally, we presented communication metaphors used by the reference implementation and a way to estimate the liveliness of team members based on the number of expected messages. The reference implementation is publicly available as open source.

Part III

General ALICA

8 Generalising ALICA

8.1 Introduction

Propositional ALICA offers a variety of modelling options to tackle multi-agent scenarios. However, the fact that it is limited to propositional semantics can cause an explosion of specific language elements, most strikingly, behaviours. In pAL-ICA, every behaviour is static. Thus, each behaviour essentially performs the same action, regardless in what situation it is called. While this limitation is somewhat lifted by the fact that behaviours are black boxes, and thus can be Turing complete programs interacting with the belief base, this is not represented within ALICA and therefore cannot be coordinated or reasoned about by means internal to ALICA. We therefore aim for a more elaborate modelling technique, where behaviours, and thus plans as well, are parametrised, such that they can be adapted to different situations. Thereby a generic behaviour $DriveTo(x)$ can express what would otherwise be expressed by multiple behaviours such as $DriveTo_{Kitchen}$, $DriveTo_{Floor}$, $DriveTo_{Door}$, and so on.

Should parameters with potentially infinite domains be required, the corresponding program is no longer expressible in pALICA. In the following, we will discuss the problem of pure propositional programs in the context of some example domains. Afterwards, we will develop a generalised version of ALICA, where plans and behaviours are parametrisable, and where reasoning in parameter spaces only bounded by memory is allowed.

8.1.1 Standard Situations

Let us consider a problem range from the RoboCup scenario. In robotic soccer, an ongoing game is often stopped due to rule violations, e.g., when the ball rolls outside of the field, or when a robot fouls another. Such a rule violation is followed by a "standard situation". In these situations, the referee positions the ball on the field, names one of the participating teams the attacker and issues a command to the robots to position themselves. After giving the robots some time to react and complete the positioning, the referee will start the game again. When the game is started, the attacking team is required to play a pass, after which the defending

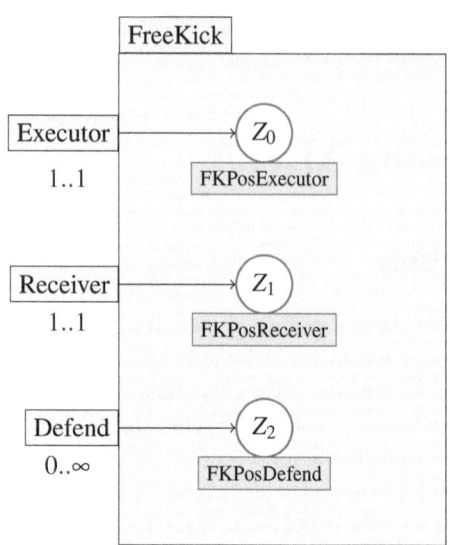

Figure 8.1: Example Standard Situation Plan in pALICA

team is allowed to intervene. The positioning phase of a standard situation is governed by various rules. For instance, no defending robot is allowed to be closer to the ball than a certain distance, e.g., 3 m. As standard situations occur quite often, a team with good positionings has a distinct advantage over one that has not. Hence, teams invest a lot of time and work into their implementations for standard situations. It is even common for a team to have specific positionings for specific opponents or environmental conditions such as inhomogeneous lighting. Given the number of different standard situations (throw-in, free kick, goal kick, etc.), and opponents, a plethora of different strategies are possible, very similar to actual soccer.

Figure 8.1 sketches the positioning phase of such a strategy implemented in pALICA. There is very little information about the actual strategy contained in this plan, it only requires one robot to act as the executor, passing the ball, one robot acting as receiver of the pass and assigns the rest of the robots to the defending task. Different strategies for the same situation would look very similar, as the actual strategy for positioning is contained in the three behaviours. While it is very easy to use and maintain several strategies for the same problem using pALICA by employing plantypes, conditions and utilities, it is impossible to express solutions to the positioning problem concisely without using several unique behaviours per strategy. This is the propositional explosion we mentioned at the beginning of this chapter. The resulting vast number of necessary behaviours can become difficult to handle and maintain during development. This is avoided by general ALICA.

We omit detailing possible implementations for the three behaviours in Figure 8.1, but extract three properties which they have in common.

Common execution: all behaviours involved are used to reach a certain, albeit different, position on the field. Given the destination, they use the same algorithms and parameters to determine motor commands (e.g., path planning and velocity controllers).

Complex decision procedure: As the destination varies over each behaviour, so does the implementation that calculates it. The resulting algorithm can become fairly complex and is susceptible to implementation errors.

Interaction: The behaviours interact with each other in that the positions depend on each other. For instance, the executor, ball, and receiver form a line, along which the pass is played later on.

We deem these properties to be general, that is, they occur in other scenarios as well, where a number of agents need to coordinate first-order entities (in this case, their positions) dynamically and with respect to the environment.

From the perspective of planning and machine learning approaches, the structure depicted in Figure 8.1 is also less than ideal. Both planning and learning are typical solved by searching a hypotheses space, e.g., the space of all programs (for instance in [96]) or task-graphs (as in [106]) for a solution to their input problem. One of the main characteristics of the search space are the behaviours in question. For the positioning problems, the search space is reduced to the possible combinations of available behaviours, which features little structure, and little information usable to guide a search.

In contrast, a first-order approach can drastically reduce the amount of behaviours needed. In the particular case of the positioning problem, only a single behaviour would be needed, which corresponds to the common execution algorithms mentioned above. Furthermore, as a first-order approach removes the need for complex black boxes, it allows planning and learning algorithms to search within a far richer representation of possible solutions to the problem in question.

One machine learning research area able to deal with this kind of search space is *Inductive Logic Programming* (ILP). An in-depth introduction into ILP is given by Nienhuys-Cheng and Wolf [111]. First order planning is tackled in various ways, such as by Regression Planning within the Situation Calculus [134], or the

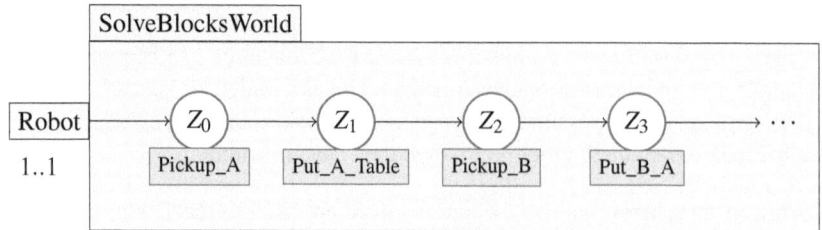

Figure 8.2: A STRIPS Action Sequence in pALICA

Relational Bellman Algorithm by Kersting et al. [84]. In the following, we refer to any algorithm that generates plans by planning, learning, or otherwise as a
generative algorithm.

8.1.2 Blocks World

As another motivating example, consider the blocks world, a prominent AI toy
problem, which was originally described by Winograd [179]. In the blocks world,
an agent or robot is tasked with building predefined structures out of geometric
primitives. Nilsson [113] introduced an elementary version, which limits the geometric primitives to cubical blocks and a large table, holding all blocks. Each
block is said to be on a single other object, either the table itself, or another block.
Further, each block is either clear, or there is a single other block on it. The robot
is able to pick up a clear block if it does not carry block, and put a block it carries
on the table or on another clear block. This elementary blocks world is maybe
the most well-known example for STRIPS planning.[1] This even led to some real-
world experiments, such as the one by Toussaint et al. [169]. In the following, we
refer to this elementary blocks world simply as blocks world.

The solution to a blocks world problem is a sequence of actions of the form
$Pickup(A), Put(A, Table), Pickup(B), Put(B, A), \dots$. As STRIPS plans are propositional, these actions are actually pretty-printed constants, which poses no theoretical problems, since the number of blocks is finite. Representing such a plan
as a pALICA program is straight-forward, each action would correspond to a behaviour within a state, and the states are connected by transitions reacting on the
successful completion of the corresponding action (see Figure 8.2).

---

[1]   Listing articles, books, and lectures that use the blocks world domain in conjunction with a
      STRIPS planner would go far beyond the scope of this thesis. At the time of writing, a google
      search for "strips blocks world" returned over $3.2 \cdot 10^7$ results.

However, targeting the problem in such a way would potentially require $n^2 + n$ behaviours in a blocks world with n blocks. Generating this amount of behaviours seems very inefficient, especially since most behaviours would do almost the same, namely either picking up a block or putting one on something else. Thus, the problem of executing a blocks world plan features the same common execution property as the standard situation scenario, however there is no interaction, as only one robot executes the plan and there are no further decisions to be made, as a static and deterministic plan is provided.

In the following, we will introduce a generalised version of ALICA based on pALICA as introduced in Part II. We will show that the general version is strictly more expressive while retaining its ability to dynamically adapt to changing situations.

8.2 Behaviour Parameters and Plan Variables

The fact that multiple behaviours share the same execution pattern can be tackled by introducing parameterised behaviours similar to first-order actions in languages such as IndiGolog [57] or FLUX [167].

Definition 8.1. Each behaviour b has a potentially empty list of variables $\vec{x}$, its parameters. We write $b(\vec{x})$ to indicate behaviour b with its parameters $\vec{x}$.

By allowing behaviours to feature parameters, the problem of common execution patterns can be dealt with. Each occurrence of a behaviour in the program can thereby have a different set of parameters. This allows the blocks world plan to be described using only two behaviours, $Pickup(x)$ and $Put(x,y)$. However, even though infinitely many different behaviour instances[1] can now be expressed using a finite amount of behaviours, and thereby a finite amount of code, this is not sufficient to deal with parameters whose values are unknown or only partially known before runtime. This is the case in the standard situation scenario above.

Consider the following simple example: A robot is tasked with finding a book in an office, and identifying it by moving towards it. Before runtime, the specific book the robot will find is unknown, as there are various different books in the office. The concrete book the robot should move towards only becomes apparent after the robot has searched for some time and the corresponding beliefs are inserted into its belief base. Figure 8.3 shows how a corresponding plan could look like. The robot executes the behaviour *LookAround* until a book is represented

---

[1]   or, more precisely, up to a memory-bounded number of instances.

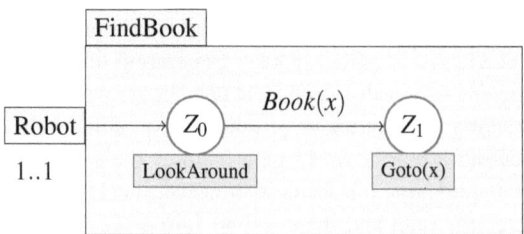

Figure 8.3: Example Plan: Finding a Book

in the belief base, and that very object is passed as parameter to the behaviour $Goto(x)$.

However, the plan *FindBook* is not expressible in pALICA, even when allowing behaviour to have parameters. What is needed additionally is a link between free variables in conditions and behaviour parameters. This allows the reasoning task, in this case the identification of a book to be defined within the plan. Similarly, this allows the identification of positions in the standard situation scenario to be defined in the plan as well.

In the following, we extend plans, states and plantypes, i.e., the elements which structure an ALICA program, with the appropriate notions to allow variables to occur in plans and relate them to each other.

Definition 8.2. Each plan p has a potentially empty list of unique variables $\vec{x}$, written $p(\vec{x})$. Each plantype P has a potentially empty list of variables $\vec{x}$, written $P(\vec{x})$.

Definition 8.3 (Independence of Plans). All plans, plantypes, and behaviours are standardised apart.

Thus, each plan and plantype has a set of unique variables available, which can be referred to by conditions. According to Definition 8.3, different plans and plantypes are associated with pairwise disjoint sets of variables. In order to relate variables of different plans to each other, we introduce the notion of bindings.

Definition 8.4. Each state s within each plan $p(\vec{x})$ defines a possibly empty substitution $\theta(s)$, called a *binding*, such that $\theta(s) = \{y_1 \mapsto x_1, \ldots, y_n \mapsto x_n\}$, all x_i are terms with variables only among the variables of the corresponding plan, $(\forall x_i) \text{vars}(x_i) \subseteq \vec{x}$, and all y_i are variables of behaviours or plantypes in s, $(\forall y_i) y_i \in \bigcup_{b(\vec{y}) \in \text{Behaviours}(s) \vee P(\vec{y}) \in \text{PlanTypes}(s)} \vec{y}$.

Definition 8.5. Each plantype $P(\vec{x})$ defines a possibly empty substitution $\theta(P)$, called a *binding*, such that $\theta(P) = \{y_1 \mapsto x_1, \ldots, y_n \mapsto x_n\}$, all x_i are terms with variables only among the variables of $P(\vec{x})$, $(\forall x_i) \operatorname{vars}(x_i) \subseteq \vec{x}$, and all y_i are variables of plans in P, $(\forall y_i) y_i \in \bigcup_{p(\vec{y}) \in P(\vec{x})} \vec{y}$.

Thus, each state and each plantype declares a local parametrisation of the plans, plantypes, and behaviours it contains, allowing to relate variables with each other, but also allowing for static substitutions. Note that bindings do not necessarily bind all variables of sub-plans or behaviours, it is entirely possible leave sub-variables unbound. Similarly, variables of the parent plan can occur arbitrarily often in a binding.

Variables of plans can now serve as means to express dynamic bindings through conditions within plans. For instance, a condition attached to a transition can refer to a book just found and bind it to a plan variable, which in turn is bound to a behaviour variable by a state. This can easily be achieved by allowing conditions to have free variables among the variables of their plan. However, in some scenarios, one might not want to identify a unique ground value for each variable. For instance in the standard situation scenario in above, calculating precise positions is cumbersome, instead one might only want to list a set of properties such positions should fulfil. Furthermore, as bindings can be used to pass variables down the plan hierarchy, such that they are eventually used as behaviour parameters, one might only want to constrain the values for each variable to values suitable to the problem tackled at each level, and leave the details to sub-plans. In this particular scenario, a plan at a high level can describe the general problem constraints, such as the set of game rules the robots must observe. It can then make use of strategy specific sub-plans, bundled in a plantype. Each of these sub-plans can further constrain the variables according to a specific strategy, such as an aggressive, yet risky strategy for free-kicks in the opponent's half, or a defensive strategy in case the team is in the lead.

This idea leads us to a constraint programming approach, where each individual plan does not ground parameters passed to behaviours, but states requirements, or constraints, ground solutions must fulfil. This way, the resulting behaviour of the multi-agent team can be expressed in a declarative manner, decoupled from implementation details of the underlying solver. Furthermore, a generative algorithm does not need to concern itself with the question whether or not generated conditions ground all variables passed to behaviours in all circumstances.

Languages such as FLUX or GOLOG [98] enforce that each action is fully grounded when posted for execution. This is vital, as there is no further reasoning happening before a corresponding action is executed, and the result of executing a non-ground action such as $Goto(x)$ is not defined. In contrast, we allow for non-ground constrained parameters, which can be grounded to appropriate values by issuing a query to a constraint solver on-demand, thus solving the problem of non-ground actions by an intermediate reasoning step. In Section 8.4, we will formally introduce the notion of a constraint formula in the context of ALICA. In Chapter 9, we will discuss means to solve the resulting constraint problems and thereby dynamically provide ground values.

8.3 Agent Variables

While the notion of plan variables potentially allows for expressive descriptions of cooperation using first-order terms, it turns out that in some instances, this is not sufficient to capture the intended team behaviour in a concise manner.

Consider the popular Foraging Scenario, used for instance by Campbell and Wu [21]. Here a team, or a swarm of agents [91] is tasked with searching and retrieving certain items, such as food or resources, while at the same time protecting their base or nest. The scenario has mainly been used to investigate role or task allocation techniques, which dynamically decide for each individual agent whether it should forage or protect. In this case, the precise number of agents assigned to a task is unknown until runtime, only an upper bound is given, even the total number of agents may be unknown until runtime. Further, the number of agents foraging or protecting can change dynamically.

Given a suitable role- and task allocation approach, the individual behaviours need to be specified. So far, expressing these behaviours using constraints in ALICA requires plan variables. Each plan variable would represent an agent's target position given it is tasked with foraging or protecting, respectively, thus twice as many variables are needed as agents can possibly participate. The resulting constraint problem would be overly complicated in all situation, as each agent only takes on one task at a time, so at most half the variables would be actually used. Furthermore, in a realistic scenario, the number of agents actually active would almost always be lower than the number of possibly active agents.

Example 8.1. *Consider the following simple example constraints for the Foraging Scenario:*

$$(\forall a, b \in \mathcal{A})MinimalDist(a, b, \varepsilon) \vee a = b \tag{8.1}$$

$$(\forall a)\, In(a, P, Protecting, z) \rightarrow Near(a, Base) \tag{8.2}$$

$$(\forall a, b)a \neq b \wedge In(a, P, Foraging, Search) \wedge In(b, P, Foraging, Search) \tag{8.3}$$
$$\rightarrow MinimalDist(a, b, SearchRadius)$$

The constraints state that all agents should maintain a minimal distance to each other (8.1), agents assigned to the task Protecting should stay near the base (8.2), and agents assigned to the task Foraging, which inhabit the state Search should maintain a larger distance to each other in order to search more efficiently (8.3). The domain-specific predicates Near and MinimalDist can be expressed as:

$$Near(a, l) \stackrel{def}{=} (\exists p)Pos(a, p) \wedge \sqrt{(p.x - l.x)^2 + (p.y - l.y)^2} < 100$$

$$MinimalDist(a, b, d) \stackrel{def}{=} (\exists p_1, p_2)Pos(a, p_1) \wedge Pos(b, p_2) \wedge$$
$$\sqrt{(p_1.x - p_2.x)^2 + (p_1.y - p_2.y)^2} > d$$

This concise way of formulating the constraints makes use of a domain-specific predicate within $\mathcal{L}(Pred, Func)$, *Pos*, and implicitly assumes it is functional in the second parameter. Such predicates are similar to functional fluents in the Situation Calculus [134], however, they do not express the state of the world or knowledge about it, but represent intended values, or rather allow constraints to assert *intentions* or *obligations*. Thereby, such constraints can be seen as a declarative extension of plans, which otherwise formulate intentions in a procedural way.

In order to support the specification of constraints which abstract from the actual agents that execute a plan or a task at a specific point in time, a set of corresponding axioms is needed to fill in the gap between constraints and beliefs:

Definition 8.6. Let F be a finite set of binary predicate symbols, and for every $P \in F$, let Σ_A contain

$$(\forall a \in \mathcal{A})\,((\exists x)P(a, x)) \wedge ((\forall x, y)P(a, x) \wedge P(a, y) \rightarrow x = y)$$

Thereby, all elements of F are functional in their second argument, given the first is an agent. We extend the belief axioms of pALICA, Σ_b (Definition 5.6) to include Σ_A. Predicates constrained by beliefs in this form are called *functional agent fluents*.

Thus, each formula in Σ_A enforces the existence of a single value x per agent a such that $P(a,x)$ holds. In other words, $P(a,x)$ attaches a variable to agent a under the name P. A domain-specific extension to Σ_A in Σ_{dom} can enable similar constructs for other domain-specific elements agents are able to modify and reason about. In the next section, we will clarify how constraints of this form can be incorporated into plans.

8.4 Constraints in ALICA

In the previous section, we already introduced some examples for constraint formulae. Intuitively, a constraint is a formula which in some way limits valid choices for domain entities. Formally, this is described by the following definition:

Definition 8.7. A constraint is a formula $\phi \in \mathcal{L}$ with free variables $\vec{x} = x_1, \ldots, x_n$, $n \geq 0$. ϕ is said to *constrain* $\vec{x}$. Every x_i is representing a value in a domain D_i. A *solution* to a constraint formula with respect to a belief base B is a substitution θ such that $\vec{x}\theta$ is ground, $\phi\theta$ is consistent with B, and $\vec{x}\theta \in D_1 \times \ldots \times D_n$.

By Definition 5.9, $B \vdash^a_{\mathcal{L},\theta} \phi$ denotes that agent a proves ϕ with respect to belief base B in logic $\mathcal{L}$ and θ is the computed answer, which constitutes a solution for the queried formula ϕ.

Definition 8.7 captures the most general notion of a constraint satisfaction problem according to Section 2.4. In the following, we refrain from explicitly stating domains in the interest of conciseness. The constraints stated in Example 8.1 do not fit into this definition, since they contain quantified variables, for which values are sought. As the corresponding sorts are finite, this is merely a technical problem.[1] Hence, formulae such as those used in Example 8.1 can be unfolded into corresponding constraints.

Definition 8.8. We denote the unfolded constraint of ϕ as $\Gamma(\phi)$. $\Gamma(\phi)$ can be obtained from ϕ by recursively replacing

- all occurrences of $((\forall a)\psi)$ by $\bigwedge_{A \in \mathcal{A}}(\psi\{a \mapsto A\})$ if $\text{In}(a,p,\tau,z)$ or $\text{Succeeded}(a,p,\tau)$ occurs in ψ for some p,τ,z,

- all occurrences $((\exists a)\psi)$ by $\bigvee_{A \in \mathcal{A}}(\psi\{a \mapsto A\})$ if $\text{In}(a,p,\tau,z)$ or $\text{Succeeded}(a,p,\tau)$ occurs in ψ for some p,τ,z,

---

[1]   While we do not use many-sorted logic (cf. [19, 52]) to define ALICA, the principle is the same.

- all occurrences of $((\forall x)\psi)$ by $\psi\{x \mapsto x_i\}$ where x_i is a variable unique to P and A if $P(A,x)$ occurs in ψ, P is a functional agent fluent according to Σ_A, and $A \in \mathcal{A}$,

- all occurrences of $((\exists x)\psi)$ by $\psi\{x \mapsto x_i\}$ where x_i is a variable unique to P and A if $P(A,x)$ occurs in ψ, P is a functional agent fluent according to Σ_A, and $A \in \mathcal{A}$.

The conditions referring to the predicates In and Succeeded in this quantifier elimination step are used to ensure that only variables ranging over agents are replaced. It is easy to see that constraint unfolding terminates, as with each operation a quantifier is removed from the formula. Note also that unfolding can be done as soon as $\mathcal{A}$ is known, i.e., during the initialisation phase of an ALICA agent, as $\mathcal{A}$ is considered to be static throughout runtime.

A formulae ϕ occurring in plan p is *applicable* for unfolding, if it is plan local to p, and each variable of ϕ occurs at most once as second parameter in a functional agent fluent. Note that the empty conjunction is treated equivalently to $\top$ and the empty disjunction equivalently to $\bot$, following the standard semantics.

Consider the following additional constraint in the Foraging scenario:

$$(\forall a,x)\,\mathrm{In}(a,P,Foraging,Search) \rightarrow Pos(a,x) \wedge Unexplored(x) \qquad (8.4)$$

Intuitively, this constraint forces robots which are currently searching to move towards unexplored space by constraining the robots' position variables using the background predicate *Unexplored*, which possibly refers to some map represented in the robots' belief states. As the robots move around, previously unexplored territory becomes known, thus the interpretation of *Unexplored* changes dynamically. Especially when a Foraging robot explores the fringe of the known territory, the map is updated constantly and thus in each iteration previously unexplored positions become explored, hence solutions to the constraint satisfaction problem need to be updated constantly. Semantically, this dependency on the current belief is captured by Definition 8.7, since a solution to a constraint problem always relates to a belief base.

Constraints in ALICA are meant to control the agents' behaviour, as such they can occur in plan conditions, e.g., as part of preconditions or as part of conditions guarding transitions. Constraints therefore relate to achievement goals in AgentSpeak [129], as they describe properties of the world the agents should pursue. In order to distinguish between condition and constraint, we introduce a new binary operator between formulae, which is not part of the underlying language $\mathcal{L}(Pred,Func)$.

Definition 8.9. If ϕ is a sentence in $\mathcal{L}$ and ψ a constraint in $\mathcal{L}$, $\phi \rightsquigarrow \psi$ is a *guarded constraint* in $\mathcal{L}$.

Intuitively, $\phi \rightsquigarrow \psi$ (read: ϕ guards ψ) has the intended meaning that if ϕ holds, the constraint ψ should hold. This syntax and semantics are loosely based on Constraint Handling Rules by Frühwirth [49, 50]. In the following, we will formalise this notion and in Section 8.6 discuss how guarded constraints are handled during runtime.

From a certain point of view, such a construct seems unnecessary. Firstly, the set of constraints is closed under conjunctive (and disjunctive) addition of sentences (or constraints). Secondly, if no solution exists for a given constraint, the corresponding condition should evaluate to false. However, treating constraints equivalent to conditions would be short sighted. As discussed in Section 2.4, depending on the expressiveness of $\mathcal{L}$, the complexity of identifying a solution can vary greatly, e.g., from tractable problems, such as 2SAT [89], NP-hard problems such as general constraints over finite domains, to undecidable problems such as arbitrary non-linear inequalities over the real numbers. Furthermore, while finding a solution to a CSP is not harder than proving the existence of one [27], this not the case for the problem of finding an optimal solution according to some objective function. In Section 9.6, we will introduce constraint optimisation within ALICA, which allows comparison of different solutions by means of an objective function.

Thus, finding a solution to a constraint can take a significant amount of time. Moreover, even though the problem class might be decidable, solvers might sacrifice completeness for efficiency (e.g., local searches such as WalkSAT by Selman et al. [147]). Depending on expressiveness of the constraint language, the solver employed, and the scenario at hand, one might chose different semantics for the guarding operator $\rightsquigarrow$.

We can distinguish two possible interpretations for the guarding operator, a *weak* and a *strong* one. In case constraint solving in $\mathcal{L}$ is tractable, as well as for NP-hard constraint satisfaction problems of limited size, a strong interpretation, requiring the existence of a solution can be used. In cases where constraint solving is hard, or the employed solver is incomplete, a weak interpretation is more sensible.

Definition 8.10 (Weak guarding interpretation). A guarded constraint $\phi \rightsquigarrow \psi$ holds under weak interpretation if and only if ϕ holds.

Hence, under weak interpretation, an agent evaluating a guarded constraint proves the guard and asserts the constraint.

Definition 8.11 (Strong guarding interpretation). A guarded constraint $\phi \rightsquigarrow \psi$ holds under strong interpretation if and only if it holds under weak interpretation and $\psi\theta$ holds for some substitution θ grounding all free variables in ψ with respect to the agent's current configuration.

Under strong interpretation an agent is also required to prove the existence of a solution to the constraint in order to evaluate the guarded constraint. However, as constraints potentially interrelate through their free variables, such a proof is only meaningful in a context, which is provided by the agent configuration. In Section 8.5 we will discuss how this context is represented and in Section 8.7 describe how context and constraint formula yield a constraint satisfaction problem.

The difference between weak guarding and strong guarding interpretation becomes apparent when guarded constraints are attached to plans. Under weak guarding, an agent will follow a transition, if it evaluates the guard to true. Under strong guarding, it must additionally prove the existence of a solution to the constraint. In both cases, the constraint will become active afterwards, that is influence the agent's behaviour. Similarly, if no solution can be found for a constraint in a runtime condition, the corresponding plan must be aborted, while the participating agents will continue to execute it under weak guarding interpretation as long as the guard holds. In this case, even though temporarily no solution can be found, the constraint continues to influence the agents' behaviour.

Ideally, guarded constraints would be formulated in such a way that the guard entails the existence of a solution to the constraint problem, while being easy to prove, in which case strong and weak interpretation become equivalent. While it is possible to mechanically construct an overly general approximation of such a guard, using some subsumption model, such as Buntime's generalised θ-Subsumption [17], such methods are out of scope of this work.

In order to incorporate constraints into ALICA, the relevant formulae are extended from sentences in $\mathcal{L}$ to guarded constraints.

Definition 8.12. If $p(\vec{x})$ is a valid ALICA plan then its precondition, runtime condition, and all formulae occurring in transitions are guarded constraints $\phi \rightsquigarrow \psi$, such that ϕ is a sentence and ψ a formula applicable for unfolding, such that all free variables of $\Gamma(\psi)$ occur as second parameters of functional agent fluents or are among $\vec{x}$. Furthermore, the guards of the pre- and the runtime condition are plan local to p.

Post conditions describe the goal a behaviour or plan is meant to achieve. Once an agent achieves that goal, the corresponding plan or behaviour is typically terminated. Hence, post conditions do not involve constraints. Behaviours are extended similarly such that preconditions and runtime conditions involve constraints:

Definition 8.13. If $b(\vec{x})$ is a valid ALICA behaviour then its precondition and runtime condition are guarded constraints $\phi \rightsquigarrow \psi$, such that ϕ is a sentence and ψ a formula applicable for unfolding, such that all free variables of $\Gamma(\psi)$ occur as second parameters of functional agent fluents or are among $\vec{x}$.

Note that behaviours are still embedded in canonical behaviour plans, as discussed in Section 5.7.

8.5 Constraint Store

Constraint logic programming systems typically feature what is referred to as a constraint store, which on the operational level can be seen as a store for encountered and thus enforced constraints. Once new information is presented, relevant constraints are "woken" and information is propagated according to the constraint handling rules. Should the set of stored constraints be found to be inconsistent, the system backtracks [2].

From a more declarative point of view, constraint stores can be seen as generalisations of classical substitutions, since constraints express information about possible valuations and any substitution can be expressed as a set of constraints.

In ALICA, each agent maintains a constraint store, which holds the set of constraints the agent currently considers to be active. As constraints occur in the context of plans or behaviours, each constraint within the store is annotated by the respective plan or behaviour.

Definition 8.14. A constraint store is a possibly empty set of tuples (P, ϕ), where P is a set of plans and ϕ a constraint.

In classical constraint programming systems, the constraint store is modified by simplification or propagation rules. For instance, a constraint that is provably satisfied in all cases is removed, or two constraints are merged into a single one. In ALICA, constraints can be added or removed from the constraint store as the agent traverses the plan-tree. Thus, at each point in time, the agent must be able to remove a formerly asserted constraint. This is achieved by annotating each constraint with the set of plans it stemmed from. A constraint ϕ, asserted by plan p, is inserted into the constraint store as $(\{p\}, \phi)$. This allows for the identification of constraints given their plan. In the base case, this set contains a single plan. In Section 8.6.2, we will introduce an operational rule that reflects constraint propagation according to Frühwirth [50], which results in sets with multiple elements.

We extend the notion of an agent configuration from Definition 5.1 to include a constraint store.

Definition 8.15 (Agent Configuration). An agent configuration in ALICA is a tuple (B, Υ, E, R, G), such that

- (B, Υ, E, R) is an agent configuration in pALICA and

- G is a constraint store.

For any agent $a \in \mathcal{A}$, Conf(a) denotes its configuration. The initial agent configuration contains an empty constraint store, i.e., has the form $(B, \emptyset, \emptyset, \emptyset, \emptyset)$, such that B does not make any assumptions about the state of the team. In particular, it does not contain any belief of the form In(a, p, τ, z).

8.6 Rules

Similar to pALICA, an ALICA agent's internal state is transformed according to a set of operational rules. In contrast to pALICA, these rules now also involve a constraint store, which contains the set of active constraints or intentions. In Section 8.6.1, we lift the original rule system introduced in Section 5.13 to the general case and describe how and when the constraint store is updated. Afterwards in Section 8.6.2, we discuss a possible extension, where constraints are updated according to Constraint Handling Rules [50].

8.6.1 Lifting Propositional ALICA Rules

The rules system described in Section 5.13 is almost directly applicable to general ALICA. We only need to reflect the new element of an agent configuration, the constraint store, which contains the set of constraints that are active, i.e., that currently can influence an agent's behaviour. In the following we define each ALICA rule with respect to the respective pALICA rule. Most changes are minor and only affect the constraint store. We write

$$T \in S : \frac{(B, \Upsilon, E, R) \xrightarrow{T} (B', \Upsilon', E', R')}{(B, \Upsilon, E, R, G) \longrightarrow (B', \Upsilon', E', R', G')}$$

to denote that an agent with configuration (B, Υ, E, R, G) updates its configuration to $(B', \Upsilon', E', R', G')$ if it would update (B, Υ, E, R) to (B', Υ', E', R') by applying pALICA rule $T \in S$. For most rules, it is sufficient to retract constraints referring to plans that are no longer in execution. We denote a limitation of a constraint store G with respect to a plan base Υ as $C(G, \Upsilon)$:

$$C(G, \Upsilon) \overset{def}{=} \{(S, \psi) \mid (S, \psi) \in G \wedge (\forall p \in S)(p, \tau, z) \in \Upsilon\}$$

In other cases, i.e., when a plan is entered using a rule, the constraints of the corresponding pre- and runtime conditions must be asserted. For notational simplicity, we use the macro $D(G,\Upsilon)$, which retracts all constraints of plans no longer executed and asserts all constraints of pre- and runtime conditions of plans referred to in Υ:

$$D(G,\Upsilon) \stackrel{def}{=} C(G,\Upsilon) \cup \{(\{p\},\psi) \mid (p,\tau,z) \in \Upsilon \wedge$$
$$\mathrm{Pre}(p) = (\phi \rightsquigarrow \psi) \vee \mathrm{Run}(p) = (\phi \rightsquigarrow \psi)\}$$

Rules without an effect on the constraint store: The rules *Sense*, *BSuccess*, *TSuccess*, *BAbort*, *BRedo*, *BProp*, and *RoleAlloc*, as well as all rules for conflict detection and resolution with the exception of Cmd_{cr}, do not modify the plan base. Therefore, they do not modify the constraint store either.

$$T \in S : \cfrac{(B,\Upsilon,E,\mathrm{R}) \stackrel{T}{\rightarrow} (B',\Upsilon',E',\mathrm{R}')}{(B,\Upsilon,E,\mathrm{R},G) \longrightarrow (B',\Upsilon',E',\mathrm{R}',G)}$$

where

$$S = \{\,Sense, BSuccess, TSuccess, BAbort, BRedo, BProp, RoleAlloc, Sense_{cd},$$
$$Adapt_{cd}, Reset_{cd}, Auth_{cr}, TakeAuth_{cr}, DropAuth_{cr}, DropCmd_{cr}, Reset_{cr}\}$$

Rules limiting the constraint store: The rules *PAbort*, *PRedo*, *PReplace*, *PProp* and *NExpand* remove triples from the plan base. Hence, their lifted versions limit the constraint store to constraints referring to plans still active.

$$T \in S : \cfrac{(B,\Upsilon,E,\mathrm{R}) \stackrel{T}{\rightarrow} (B',\Upsilon',E',\mathrm{R}')}{(B,\Upsilon,E,\mathrm{R},G) \longrightarrow (B',\Upsilon',E',\mathrm{R}',C(G,\Upsilon'))}$$

where
$$S = \{PAbort, PRedo, PReplace, PProp, NExpand\}$$

Rules asserting and retracting constraints: The rules *Alloc*, *Adapt*, and Cmd_{cr} modify the plan base by addition and removal of triples. This is reflected in the constraint store by limitation to constraint referring only to plans still in execution and the assertion of constraints from the respective pre- and runtime conditions of newly entered plans. Note that the three rules only add triples referring to initial states.

$$T \in S : \cfrac{(B,\Upsilon,E,\mathrm{R}) \stackrel{T}{\rightarrow} (B',\Upsilon',E',\mathrm{R}')}{(B,\Upsilon,E,\mathrm{R},G) \longrightarrow (B',\Upsilon',E',\mathrm{R}',D(G,\Upsilon'))}$$

where

$$S = \{Alloc, Adapt, Cmd_{cr}\}$$

Rules clearing the constraint store: An agent initialises its configuration using the rule *Init*. In case of a failure that occurs in or is propagated to the top-level plan, the rule *PTopFail* is used to handle the failure and trigger a reinitialisation. In both cases, the constraint store is emptied, freeing the agent from any previous intention expressed as constraints.

$$T \in \{Init, PTopFail\} : \frac{(B, \Upsilon, E, R) \xrightarrow{T} (B', \Upsilon', E', R')}{(B, \Upsilon, E, R, G) \longrightarrow (B', \Upsilon', E', R', \emptyset)}$$

This leaves two special cases, namely the rules dealing with transitions.

The Transition Rule: By applying the Transition Rule, an agent traverses a transition in a state machine. This causes the corresponding constraint to be asserted.

$$Trans : \frac{(B, \Upsilon, E, R) \xrightarrow{Trans} (B', \Upsilon', E', R')}{(B, \Upsilon, E, R, G) \longrightarrow (B', \Upsilon', E', R', C(G, \Upsilon') \cup \{(\{p\}, \psi)\})}$$

where ψ is the constraint of the transition $(z_1, z_2, \phi \rightsquigarrow \psi)$ the agent traverses and p the plan which contains the transition. Thus, by entering a new state via a transition, the agent inserts the corresponding constraint into its constraint store. Note that the constraint can trivially be $\top$, since any sentence ϕ can be rewritten as an equivalent guarded constraint $\phi \rightsquigarrow \top$. Additionally, constraints referring to plans no longer executed are removed.

The Synchronised Transition Rule: Analogously to the Transition Rule, the Synchronised Transition rule inserts constraints into the constraint store and removes constraints referring to the branch of plans left by the agent. However, since multiple agents are involved in the synchronisations, the constraints of every transition belonging to the synchronisation are inserted.

$$STrans : \frac{(B, \Upsilon, E, R) \xrightarrow{STrans} (B', \Upsilon', E', R')}{(B, \Upsilon, E, R, G) \longrightarrow (B', \Upsilon', E', R', C(G, \Upsilon') \cup F)}$$

such that if s is the synchronisation in question and p the plan in which s occurs then $F = \{(\{p\}, \psi) \mid (z_1, z_2, \phi \rightsquigarrow \psi) \in s\}$.

This extension of the rule set to incorporate the constraint store is consistent with the properties discussed in Section 5.14. That is, the plan base stays consistent with the plan base axioms, no orphaned behaviour is introduced and the agent keeps believing in what it is intending. This is simply due to the fact that lifting only causes the rules to handle the constraint store. The handling of the other components of an agent configuration remains the same.

8.6.2 Constraint Handling Rules

Frühwirth [50] defines a declarative programming system based on Constraint Handling Rules (CHRs). While such rules are not in the focus of this work, it is possible to augment an ALICA engine with a set of constraint handling rules in order to simplify subsequent reasoning steps such as solving a query for specific variables. Currently, this is not supported by the implementation. Here, we briefly illustrate how such an extension can be formulated.

The following definition by Frühwirth [50] captures the syntax of CHR programs.

Definition 8.16. A CHR program is a finite set of CHRs. Each CHR is of one of the following forms:

- A *simplification* CHR is of the form: $H_1, \ldots, H_i \leftrightarrow G_1, \ldots, G_j \mid B_1, \ldots, B_k$

- A *propagation* CHR is of the form: $H_1, \ldots, H_i \rightarrow G_1, \ldots, G_j \mid B_1, \ldots, B_k$

- A *simpagation* CHR is of the form: $H_1, \ldots, H_l \setminus H_{l+1}, \ldots H_i \rightarrow G_1, \ldots, G_j \mid B_1, \ldots, B_k$

where $i > 0$, $j \geq 0$, $k \geq 0$, $l > 0$, and the multi-head $H_1, \ldots, H_i$ is a non-empty sequence of CHR constraints, the guard $G_1, \ldots, G_j$ is a sequence of built-in constraints, and the body $B_1, \ldots, B_k$ is a sequence of built-in and CHR constraints.

In the CHR framework, built-in constraints are the atomic building blocks of more complex constraints defined entirely by means of CHRs.

Intuitively, if for each constraint H_x in the head of a simplification rule an instance $H_x \sigma$ exists in the constraint store, and the corresponding guard instance $G_1 \sigma, \ldots, G_j \sigma$ of the rule can be proven with answer θ, such that θ does not substitute variables occurring in the head, then all $H_x \sigma$ are removed from the constraint store and replaced by the constraints in the body, $B_x \sigma \theta$. The propagation rule works similarly, but does not remove the head from the store. The simpagation rule is a combination of the two, $H_1 \sigma, \ldots, H_l \sigma$ are removed, while $H_{l+1} \sigma, \ldots, H_i \sigma$ are not. The resulting system is Turing complete [160].

We can provide an operational rule that incorporates propagation CHRs into ALICA. In contrast to the original CHR system, a constraint in ALICA is never removed from the store unless the agent leaves the corresponding plan. Given a set of propagation rules acting as CHR extension to ALICA, the following rule, *CProp*, defines the corresponding update to an agent configuration.

$$CProp: \frac{(B \vdash_\theta (G_1 \wedge \ldots \wedge G_j)\sigma) \wedge (S_1, H_1\sigma) \in G \wedge \ldots \wedge (S_i, H_i\sigma) \in G}{(B, \Upsilon, E, R, G) \longrightarrow (B, \Upsilon, E, R, G')}$$

where

- $H_1, \ldots, H_i \to G_1, \ldots, G_j \mid B_1, \ldots, B_k$ is a propagation rule in the CHR extension,

- θ is the computed answer to the proof of the guard.

- $G' = G \cup \{(T, B_i \sigma \theta) \mid 1 \le i \le k\}$, and

- $T = \bigcup_{k=1}^{i} S_i$.

The rule *CProp* asserts the body of a propagation rule into the constraint store if the guard holds with respect to the current set of beliefs and an instance of the head of the rule is active, i.e., is in the constraint store. Similar to rule application in CHR languages, *CProp* may be applied at most once to the same constraints to avoid nontermination. Using this rule, it is possible to transfer parts of the CHR semantics to work within ALICA.

Embedding simplification rules in this way is not possible, since the constraint store is dynamic and reflects the set of active constraints at any point in time. Consider for instance a scenario in which the two tuples $(\{p_1\}, \phi_1)$ and $(\{p_2\}, \phi_2)$ are part of the constraint store. An application of a simplification rule could remove the two tuples and add a tuple corresponding to the body of the simplification rule, e.g., $(\{p_1, p_2\}, \psi)$. If the agent now leaves p_2, it must retract this consequence from the constraint store, since constraints from p_2 are no longer relevant. Thereby, the agent also retracted the original constraint ϕ_1, which should still be active.

8.7 Queries

Thus far, constraints are maintained and managed within a constraint store, but they do not affect the behaviour of the agent. This is achieved through queries. During the runtime of the agent, behaviours, and potentially other components can query the constraint store.

We distinguish two kinds of queries, existential ones, which only require proof of the existence of a solution, and concrete ones, which demand a concrete solution. Although, as discussed earlier, these problems are equally complex, once we turn to constraint optimisation in Section 9.6 this is no longer the case, hence the distinction.

Under the strong guarding interpretation (see Definition 8.11), operational rules issue existential queries when evaluating plan-conditions.

Definition 8.17. A query q is a tuple $(c, \vec{v}, \phi)$, consisting of a context c, which is either a plan or a behaviour, in which the query is posed, a list of variables $\vec{v}$, for which values are to be calculated, and a constraint formula ϕ with free variables only among $\vec{v}$.

In order to answer a query, a corresponding constraint satisfaction problem is constructed with respect to the current agent configuration (B, Υ, E, R, G). Intuitively, this is done by traversing the plan-tree upwards from the context in which the query was posed towards the top-level plan and collecting all active constraints and bindings on the way. Constraints active in the context of children or siblings are not considered, following the locality principle. Thus, a query posed in the context of plan p might be more specific than the same query posed in the context of the parent plan of p.

The upwards traversing of the plan-tree is captured by the recursive function t in Figure 8.4. The first case captures the termination condition, when the top-level plan is reached. The second case is relevant if the context c is a behaviour or a plantype. It then applies the parent state's bindings $\theta(z)$ and adds any active constraints associated with the corresponding parent plan p. The last case handles plans, and applies the bindings of the respective plantype P.

In this way, the traversal $t(c, \vec{v}, \phi, \emptyset, \Upsilon, G)$ for the query $(c, \vec{v}, \phi)$ yields the tuple (ψ, θ), consisting of the CSP ψ to be solved, and a substitution θ, i.e., the composition of all collected bindings. The CSP is constructed as the conjunction of ϕ and all relevant active constraints from the constraint store G.

The substitution θ captures the relationship between the queried variables and the context. Each variable is mapped by θ to the one occurring highest in the plan-tree, due to the transitive application of the state and plantype bindings. This substitution will play an important role for solution tracking and coordination, as we will discuss in Section 9.5.

Subsequently, an agent can proof $B \vdash_\eta \psi$ to answer the query. In case of a concrete query, $\vec{v}\theta\eta$ is the resulting solution vector.

Example 8.2. *In a simple strategy for attacking in the RoboCup scenario, the ball possessing robot drives towards a position from which it can score. A second robot*

$$t(c,X,\phi,\theta,\Upsilon,G) = \begin{cases} (\phi,\theta) & \text{if } c = p_0 \\ t(p,X',\phi',\theta',\Upsilon,G) & \text{if } (p,\tau,z) \in \Upsilon \wedge (c \in \text{Behaviours}(z) \\ & \qquad \vee c \in \text{PlanTypes}(z)) \\ & \text{where} \\ & \bullet X' = X\theta(z) \\ & \bullet \phi' = \phi\theta(z) \wedge \\ & \quad \bigwedge_{(\exists \mathcal{P},\nu)(\mathcal{P},\psi) \in G \wedge p \in \mathcal{P} \wedge \nu \in \text{vars}(\psi) \cap X'} \psi \\ & \bullet \theta' = \theta \circ \theta(z) \\ t(P,X',\phi',\theta',\Upsilon,G) & \text{if } (c,\tau,z') \in \Upsilon \wedge P \in \text{PlanTypes}(z) \wedge \\ & \quad c \in P \\ & \text{where} \\ & \bullet X' = X\theta(P) \\ & \bullet \phi' = \phi\theta(P) \\ & \bullet \theta' = \theta \circ \theta(P) \end{cases}$$

Figure 8.4: CSP Construction: Upwards Traversal of the Plan-Tree

tries to hinder opponent players from reaching the ball, and the two remaining robots are positioned close to the own goal, such that they are able to defend should an opponent obtain the ball. Figure 8.5 depicts a corresponding ALICA plan.

The plan consists of the three respective tasks and is successfully completed once the Attacker kicks the ball towards the opponent goal (whether it scores or not). The detailed behaviour of the other robots is hidden at this level and encapsulated by the plantypes Protect(x) and DefendGoal, respectively.

The plan's semantic is largely dominated by its runtime condition ϕ, which could be defined as:

$$\phi = \phi' \rightsquigarrow \psi = ((\exists a,z)\,\text{In}(a,Attack,Attacker,z) \wedge HasBall(a)) \rightsquigarrow$$
$$\Gamma(ScorePosition(x) \wedge ((\forall a)\,\text{In}(a,Attack,Defender,Z_4) \rightarrow$$
$$(\exists defPos)TargetPos(a,defPos) \wedge$$
$$Between(OwnGoal,Ball,defPos)))$$

Hence, the plan must be aborted, if there is no robot in ball possession (HasBall(a)), which can take on the task Attacker. Further, the plan variable x, which is free in φ, is meant to indicate the position towards which the attacking robot dribbles. x is constrained by the macro ScorePosition(x), whose details are not of any importance here, but it likely takes the position of the opponents and the physical limitations of the robot's kicking device into account. Additionally, φ constraints the functional agent fluent TargetPos of all defending robots to be between the ball and the goal. Further details about the positions of the blocker and the defenders are left open.

DribbleTo(x), executed by the attacker will issue a query (DribbleTo, (x), ⊤) every iteration in order to obtain a destination, which in turn is translated into a movement command. Traversing over the plan-tree yields the CSP ⊤ ∧ ψ, which does not contain any constraints from the plantypes Protect(x) or DefendGoal. Intuitively, the details are not relevant for the attacker, it suffices to know that the defenders are positioned behind it.

A blocker on the other hand, executes the plantype Protect(x), whose realising plan would constraint the blocker's position in relation to the attacker's target position x, and by traversing, the constraint of φ is appended to any query about the blocker's position from within Protect(x). Similarly, there exists another set of constraints within the plantype DefendGoal, which defines the positions of the defenders. CSPs constructed from queries about their positions from within DefendGoal will include ψ, since ψ states requirements about them, i.e., shares variables with the query due to the functional agent fluent TargetPos.

It is important to note the difference between the actual position of an agent and its constrained target position. The actual position is, for a given moment in time, fixed and cannot be changed, while the target position, reflected by the agent fluent $TargetPos(a, p)$, can be subject to arbitrary constraints.

8.8 Summary

Motivated by the fact that many problems cannot easily be expressed using propositional ALICA, we extended this basic language with variables for plans, plantypes, behaviours, and agents. This allows for constraints to be integrated into the language as part of conditions. We discussed two different semantics of the resulting guarded constraints, namely the weak guarding and the strong guarding semantics. Weak guarding should be chosen whenever the corresponding solver is incomplete or needs a significant amount of time to prove the existence of a solution. Strong guarding requires that a constraint has a solution before it is activated.

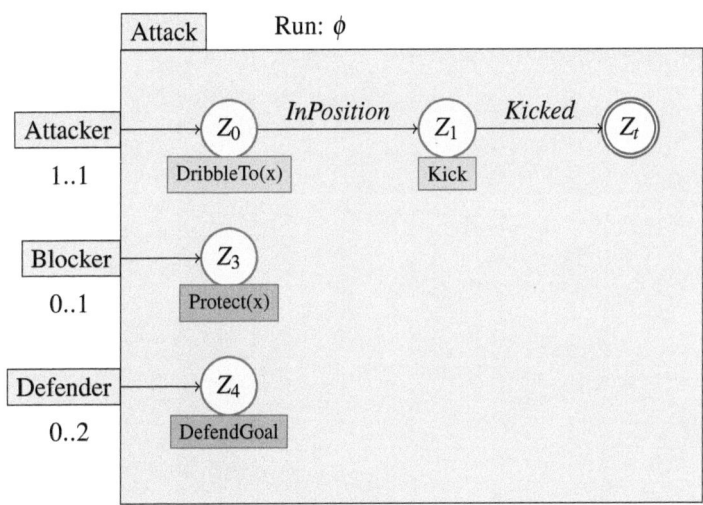

Figure 8.5: Example Plan: A Strategy for an Attack in RoboCup

The operational semantics of pALICA was lifted to this more expressive case by the integration of a constraint store into the agent configuration. The constraint store maintains the set of active constraints with respect to the plans in execution. Finally, we presented how components can query variables for values given a plan or a behaviour as context of the query. The next chapter is concerned with solving the resulting constraint satisfaction problems.

9 Solving Constraint Problems

So far we discussed how constraints are managed by an ALICA agent. The way constraints are managed and updated does not place any restriction on the solver used. It can be chosen according to the class of constraint problems formulated. However, discussing solvers for all possible, or even reasonable problem classes is out of scope of this work. In the following, we will focus on a specific class of constraint satisfaction problems, which is sufficiently expressible for a wide range of robotic problems. Firstly, we investigate the expressive need of some example domains, focussing on robotic scenarios. Based on these examples, we formally define a suitable class of constraint problems geared towards robotic and multi-robotic scenarios. Afterwards, we present a solver for this problem class and discuss extensions to it that allow for solving of constraint optimisation problems, tracking of solutions over time, and coordination of solutions within the team.

9.1 Exemplary Constraint Satisfaction Problems

Let us again consider standard situations in the robotic soccer domain. In a standard situation, the positions of up to four robots on the field need to be determined, given the ball's location and any perceived opponent position. At most, such a problem has 12 dimensions (three per robot, as a position is described by a two dimensional point and an orientation). Compared to typical constraint satisfaction problems appearing in industrial domains, that is very small.[1] However, solutions need to be found fast, under soft real-time conditions, as we will discuss in Section 9.4. Furthermore, the domain is dynamic, and hence solutions change over time.

The specific constraints needed to describe behaviour during a standard situation typically fall into one of the following categories:

linear: Enforcing robots to stay within or outside of rectangular regions,

quadratic: When distances between robots and objects are involved,

---

[1] Compare, for instance, benchmark problems used during the Satisfiability Modulo Theories Competition (SMT-COMP).

polynomial: When more complex geometric relationships are used, such as circumference tests.

Additionally, these constraint problems are not purely conjunctive, since multiple distinct solutions can be described. As an example, consider the standard situation free-kick, in which the position of two robots are constrained, such that one receives a pass (p_1), and another (p_2) tries to prevent opponents from intercepting the pass. An intuitive instruction for the team could be phrased as: "If there is an opponent able to intercept the pass, it should be blocked by p_2." For better readability we denote the position of an object o by two variables o_x and o_y. In the following, the ball is denoted by b.

We assume an opponent o can intercept the pass vector, if it is closer to the pass vector $\vec{d} = p_1 - \vec{b}$ than the ball is to the interception point. Given

$$\vec{c} = (o - b)$$

$$\vec{v} = \frac{1}{|d|} \begin{pmatrix} c_x d_x + c_y d_y \\ c_x d_y - c_y d_x \end{pmatrix}$$

This is expressed by:

$$\phi = v_x > 0 \wedge v_x < 1 \wedge v_x^2 > v_y^2$$

Therefore, if ϕ holds for an opponent, the blocker should be positioned in front of it:

$$\psi = |p_2 - v_y \vec{d}_\perp + o| < 0.05$$

Where $\vec{d}_\perp$ denotes the vector orthogonal to $\vec{d}$. Since there are multiple opponents, a variant of $\phi \rightarrow \psi$ is conjunctively added to the problem for every perceived opponent.

One approach to tackling such constraint satisfaction problems is to discretize the domain of each variable to a suitable grid (e.g., 1cm width) and solve the resulting finite domain problem. However, this yields a very large solution space, especially when multiple robot positions are considered. In this specific case, the resulting space would have about $4.7 \cdot 10^{12}$ elements ($2.2 \cdot 10^{25}$ in case of four robots).

Such an exponential effect can easily negate any advantage finite domain solvers have over continuous domain solvers. Moreover, discretization can potentially cause numerical problems with respect to specific constraints due to aliasing and other effects. Therefore, we argue that Boolean combinations of rational inequalities over the real numbers constitute the most suitable constraint system for this scenario.

In another scenario, suppose a mobile robot is tasked with fetching an item, for instance a cup of coffee from the kitchen. It is equipped with a robotic arm, whose end-effector can grab cup-sized objects. In order to accomplish its task, the robot has four behaviours available, namely $DriveTo(\vec{x})$, $PositionGripper(\vec{y})$, $Grab$, and $Release$. The behaviour $DriveTo(\vec{x})$ is able to navigate the building in order to reach a position described by $\vec{x}$, consisting of the Cartesian coordinates x and y, and an orientation α. The behaviour $PositionGripper(\vec{y})$ controls the robotic arm, such that the end-effector reaches a position described by the three-dimensional vector $\vec{y}$. For simplicity, we omit specifying the end-effector's orientation. Further, suppose the robotic arm originates at the robot's centre, and has five revolute joints. Let M_i denote the matrix which describes the transformation of the i-th joint with respect to its angular configuration β_i, the position of the end-effector in the world coordinate system can be calculated as

$$\vec{y} = M_5 M_4 M_3 M_2 M_1 M_0 (x, y, 0, 1)^T$$

where M_0 describes the rotation resulting from the robot's orientation.

Additionally to the standard situation scenario, this inverse kinematics problem requires trigonometric functions, such as sine and cosine. In the next section, we formulate a problem class suitable to express both problems.

The resulting equation has eleven variables, one for each degree of freedom, plus the end effector's position and the robot's position. Thereby, the two problems of finding $\vec{x}$ and $\vec{y}$ are related. Together with appropriate information about the cup's location in the kitchen and the layout of the kitchen, a simple constraint problem can be formulated, such that both behaviours interact in a way, that the robot moves to a suitable position from where it can grab the cup.

9.2 Non-Linear Continuous Constraint Satisfaction Problems

In [155], we discussed the problem class of non-linear continuous constraint satisfaction problems given in the following way:

Definition 9.1. A continuous non-linear constraint satisfaction problem (CNLCSP) is a triple (ϕ, X, C) consisting of:

- a propositional formula ϕ with variables P,

- a set of n variables X each ranging over $\mathbb{R}$,

- Every $p_i \in P$ identifies a constraint $c_i \in C$ such that $c_i = f_i(\vec{x}) \circ_i 0$, where $\circ_i \in \{<, >, \leq, \geq, =, \neq\}$, $\vec{x} \subseteq X$, and all f_i are arbitrary functions $\mathbb{R}^k \mapsto \mathbb{R}$.

An interpretation of a CNLCSP is a valuation function $v\colon X \mapsto \mathbb{R}$, which is extended to $P \mapsto \{\top, \bot\}$ by

$$v(p_i) = \begin{cases} \top & \text{if } v(f_i(\vec{x})) \circ_i 0 \\ \bot & \text{otherwise} \end{cases}$$

The interpretation v is a *solution* for formula ϕ if and only if ϕ, in which all variables p_i are replaced by their interpretation $v(p_i)$, evaluates to $\top$ under classical propositional interpretation.

Furthermore, in [155], we evaluated solvers tackling this problem class stemming from completely different fields of research, such as interval propagation and splitting ([2, 171]), satisfiability modulo theories ([6, 48]), and local numerical searches over a landscape defined by the constraint satisfaction problem. Satisfiability modulo theories (SMT) is a relatively new approach to solving problems of various different theories by combining a theory specific solver with a solver for Boolean Satisfiability (SAT). Thereby, such a solver transfers some of the speed of recent SAT solvers (e.g., [102]) to more expressive problem classes.

However, in our evaluation, local numerical searches outperformed all others in terms of speed on the benchmark problems, which were drawn from robotic scenarios, such as inverse kinematics of two robotic arms, path-planning, standard situations in RoboCup, and swarm behaviour. We argue that this is due to the relative simple Boolean structure of the problems evaluated, such that SMT solvers could not exploit the strengths of their SAT solving component. We will return to this hypothesis in Section 9.3.

The local searches evaluated were based on the evolutionary strategy CMA-ES (Covariance Matrix Adaptation Evolutionary Strategy) by Hansen and Ostermeier [66] and resilient propagation (Rprop) by Riedmiller and Braun [136] combined with automatic differentiation. On the implementation level, the automatic differentiation is done by a modified version of AutoDiff by Shtof [153]. The Rprop-based search performed slightly better than the evolutionary strategy.

In order to solve a CSP using only a numerical local search, a transformation needs to be applied, which produces a single function to be optimised. For this, the input formula is transformed such that negations only occur in front of atoms by subsequently applying de Morgan's law and double negation elimination. Afterwards the resulting formula is mapped to a function by a transformation T. Out of the four transformation functions tested in [155], the following performed best:

Definition 9.2. Formula transformation:

$$T(\phi \wedge \psi) = \Sigma_\wedge(T(\phi), T(\psi)), \qquad\qquad T(\phi \vee \psi) = \max(T(\phi), T(\psi)),$$
$$T(a < b) = <^*(a,b), \qquad\qquad\qquad T(a > b) = <^*(b,a),$$
$$T(\neg(a < b)) = \leq^*(b,a), \qquad\qquad T(\neg(a > b)) = \leq^*(a,b),$$
$$T(\neg(a \leq b)) = <^*(b,a), \qquad\qquad T(\neg(a \geq b)) = <^*(a,b)$$
$$T(a = b) = T(a \geq b - \varepsilon \wedge a \leq b + \varepsilon), \quad T(a \neq b) = T(a < b - \varepsilon \vee a > b + \varepsilon)$$

where

$$\Sigma_\wedge(a,b) = \begin{cases} 1 & \text{if } a = 1 \wedge b = 1 \\ \min(0,a) + \min(0,b) & \text{otherwise} \end{cases}$$

$$\frac{\delta}{\delta x_i} \Sigma_\wedge(a,b) = \frac{\delta}{\delta x_i}(a+b)$$

$$<^*(a,b) = \begin{cases} 1 & \text{if } a - b < 0 \\ b - a & \text{otherwise} \end{cases}$$

$$\frac{\delta}{\delta x_i} <^*(a,b) = \begin{cases} 0 & \text{if } a - b < 0 \\ \frac{\delta}{\delta x_i}(b-a) & \text{otherwise} \end{cases}$$

$$\frac{\delta}{\delta x_i} \max(a,b) = \begin{cases} \frac{\delta}{\delta x_i}a & \text{if } a \geq b \\ \frac{\delta}{\delta x_i}b & \text{otherwise} \end{cases}$$

The function $\leq^*$ is defined analogously to $<^*$.

This way, a result of 1 resembles a satisfied constraint, while infeasible points are mapped onto values below or equal to 0. The gradient of the resulting function encodes information about unsatisfied constraints, pointing towards potentially feasible regions. This transformation together with a gradient method such as Rprop with auto-differentiation allows a wide variety of problems to be tackled in soft real-time as shown in [155]. Figure 9.1 illustrates the resulting functions.

In realistic scenarios, bounds for every variable involved are known. For example, positions in robotic soccer cannot be arbitrarily far away from the playing field and angles range between $-\pi$ and π. Thus, for every variable in a CNLCSP, we can assume a suitable interval, which contains all relevant solutions. These bounds can either be given as annotations to constraints or be derived from constraints using interval propagation. Listing 9.1 shows the pseudo-code of a basic solver

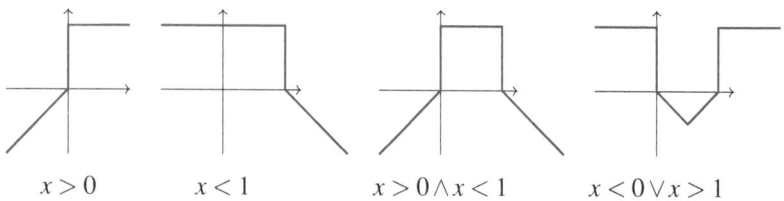

$x > 0$ $x < 1$ $x > 0 \wedge x < 1$ $x < 0 \vee x > 1$

Figure 9.1: Exemplary Transformed Constraints

following the principles discussed so far. The solver performs local searches initialised with different random points until either an upper limit of trials (maxTries) is reached or a solution is found ($f(p) > 0$). The listing also shows default values for the relevant parameters of the search, namely s_{init}, the initial step size relative to the size of each domain, inc, the factor increasing the step-size in case the sign of the gradient stays the same, dec, the factor decreasing the step-size in case the sign of the gradient changes, and $minV$ and $maxV$, the minimal and maximal step size, respectively.

However, the major drawback of applying local searches to CSPs is the inherent incompleteness. Even if the presented problem is decidable, a local search may not find a solution. More importantly, a local search cannot deduce unsatisfiability of a problem. Therefore, this solver should be used in conjunction with the weak guarding interpretation discussed earlier. In the light that these solvers are to be employed under soft-realtime conditions, this issue becomes less important. If a solution cannot be found within a certain time frame, the existence of one becomes almost irrelevant. We will discuss this topic in more detail in Section 9.4.

9.3 SMT-Solvers Revisited

Earlier, we claimed that while SMT-solvers were outperformed by local search methods on problems relevant in robotic scenarios, they should come out on top, given a sufficiently complex Boolean structure of the underlying problem. We evaluate this hypothesis using the following problem generator based on the sine generator discussed in [155], which in turn is based on the generator proposed by Shang et al. [152]:

$$g(n,l,d) = (\phi, X, C)$$

such that

- X contains d variables ranging over the reals,

```
Solve(csp) {
    // parameters : s_init = 10⁻³, inc=1.2, dec=0.5;
    // minV = 10⁻¹², maxV = 10¹⁰
    f := transform(csp); // transform the problem into a function
    let p^min and p^max be the vectors of the lower (upper) bounds, respectively;
    d := number of dimensions of the csp;
    i := 0;
    while(++i < maxTries) {
        p := random point ∈ [p₁^min, p₁^max] × ... × [p_d^min, p_d^max];
        j := 0;
        let s and h be vectors with d elements;
        set all s_k := s_init · (p_k^max − p_k^min);
        set all h_k := 0;
        while(++j < maxIterations ) {
            if (f(p) > 0) return p;
            g := ∇f(p);
            for (k:=0; k<d; k++) {
                if (g_k · h_k > 0) s_k := s_k · inc;
                if (g_k · h_k < 0) s_k := s_k · dec;
                s_k := min(max(s_k, minV), maxV);
                p_k := p_k + sgn(g_k) · s_k;
                p_k := min(max(p_k, p_k^min), p_k^max);
            }
            h := g;
        }
    }
    return FAILURE;
}
```

Listing 9.1: Rprop-Based Local Search

- C contains l inequalities of the form

$$k\Sigma_{i=1}^3 \Pi_{j=1}^3 a_{ij} \sin(2\pi x_{ij} + c_{ij}) < \theta$$

where $k = \frac{1}{\Sigma_{i=1}^3 \Pi_{j=1}^3 a_{ij}}$, all a_{ij} are uniformly distributed random values in $[-1, 1]$, all b_{ij} are uniformly distributed random values in $[-2\pi, 2\pi]$, all x_{ij} are randomly chosen variables in X, and θ is a threshold value, such that the feasible region of the constraint is approximately half the size of the whole domain, measured by random sampling. This resembles the ratio by which the solution space in pure SAT problems is divided by a single propositional variable.

- ϕ is a random 3-SAT formula containing n clauses, whose propositional variables P each uniquely identify a constraint in C.

- Satisfiability of the CSP is guaranteed in the following way: Let $\vec{s}$ be a random point in $\mathbb{R}^d$, acting as valuation function v for X. Let P' be the set of propositional variables valuated to $\top$ by the extension of v. Then, for every clause in ϕ, pick a random literal and set its sign to positive (negative) if its propositional variable is in P' (is not in P').

The SMT-solvers evaluated in [155] are not suitable for this kind of constraint system. ABsolver [6] cannot deal with trigonometric functions (which we deem important in robotic scenarios), and iSAT [48] performs poorly on this benchmark. We presume this is due to a shortcoming in its parser, which probably does not recognize reoccurring terms. However, since iSAT's source code is closed, this is pure conjecture.

Instead, we evaluate a naive SMT-solver which uses the DPLL-algorithm Chaff by Moskewicz et al. [102] to produce an assignment for the propositional variables P. Given this assignment, a purely conjunctive constraint system is constructed and solved using the local search discussed in the previous section. The local seach is restarted exactly one time. In case this does not suffice to find a solution, Chaff is called to produce the next propositional model, and the process is repeated. In case all propositional models are tried, both solvers are reset and the process starts from the beginning. In the following, we refer to this solver as Chaff+Rprop($\Sigma_\wedge$, max), indicating the algorithms and the transformation function used.

While this combination of SAT solver and local search forces the latter to solve a harder problem, since the conjunctive problem is a specialisation of the original problem, the approach can quickly pay off in two ways. Firstly, the resulting formula, which is used to evaluate individual points in the solution space, is simplified. Secondly, the local search does not have to deal with disjunctions, which can cause a misleading landscape, as discussed in [155]. Of course, in this naive approach, the two solvers are kept quite separate and there is little information exchanged between them. A more sophisticated SMT solver should be able to exchange more information, such as conflict-explanations.

Figure 9.2 shows the result for constraint satisfaction problems generated by $g(n, 50, 25)$ for varying n. The timeout for solving was set to 30 min. Constraint ratio refers to the ratio between number of clauses and literals, in this case $\frac{n}{50}$. Each point in the graph is averaged over 100 trials. All experiments were carried out single-threaded on an Intel® Core™ i7 930 CPU (2.8 GHz) running Linux 2.6.38.

Firstly, the performance of the Rprop-based search (Rprop($\Sigma_\wedge$, max)) and the CMA-ES-based search (ES($\Sigma_\wedge$, max)) follow the results shown in [155], although the problem here is slightly harder. Secondly, local searches outperform the SMT

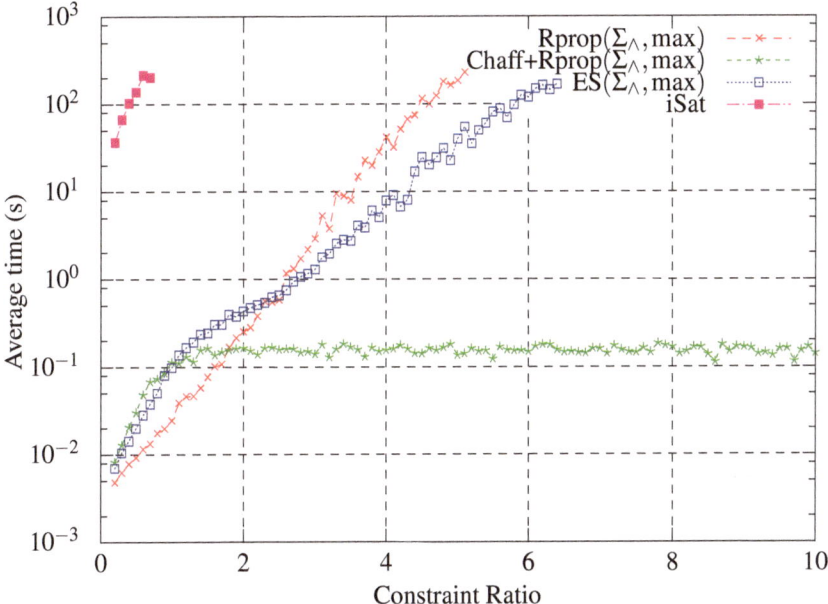

Figure 9.2: Performance of Local Searches and SMT Solvers in the 3-SAT-Sine Test Case

solver iSAT [48], evaluated here for comparison, even though the set of distinct literals is constant. Our naive SMT solver performs worse than the pure local searches for constraint ratios lower than 1, while starting to outperform the local searches at an constraint ratio of about 2. Afterwards, the performance of Chaff+Rrop$(\Sigma_\wedge, \max)$ is almost constant with respect to the constraint ratio.

We can conclude that the SAT-solving time, which should peak around a constraint ratio of 4.24 [30, 154][1], is insignificant compared to the time for processing the non-linear constraints. Secondly, for very low ratios, the SMT approach suffers from producing a harder problem for the local search than the original.

Of course, this is a highly artificial test case, which is geared towards showing the potential of SMT solving. Still, this test shows great potential if one can succeed at transferring some of this performance to real-world problems. Herein, however, lies the catch. Figure 9.3 shows the number of local search runs done by each solver in this experiment. The number of runs done by

---

[1]   after Russell and Norvig [139]

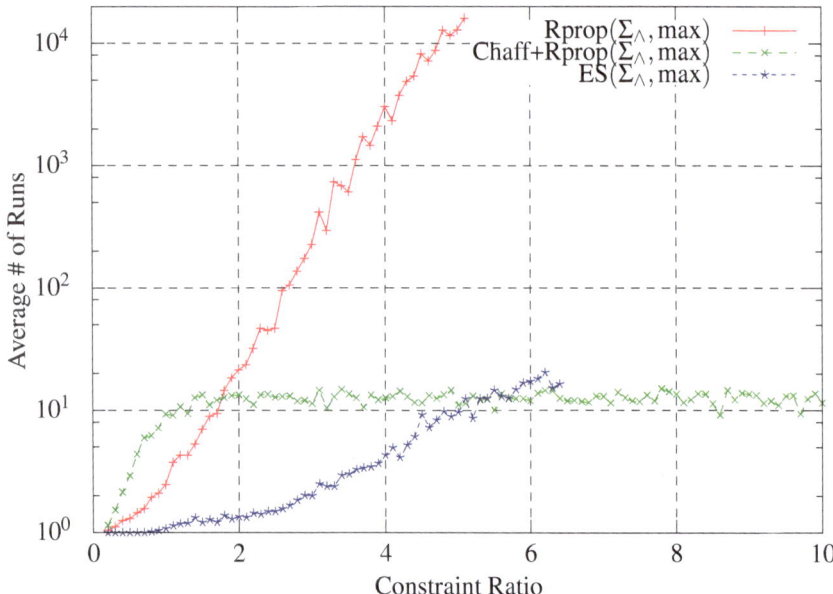

Figure 9.3: Number of (Re-)starts Performed in the 3-SAT-Sine Test Case

Chaff+Rprop($\Sigma_\wedge$, max) reaches 12.5 at a constraint ratio of around 2 and stays constant afterwards. This means that in average, only 6 to 7 propositional models were produced until a solution was found. Given that the total number of propositional models should be much higher for a constraint ratio between 2 and 4, this indicates that for any combination of sine-based literals as generated by $g(n, l, d)$, there is a high probability that a solution exists.

In more practical scenarios, this might not be the case, and, if the Boolean structure has many models, this naive SMT approach can quickly become infeasible. Such a scenario is sketched in the following example.

Example 9.1. *In robotic soccer, a suitable strategy for countering an opponent free-kick which uses four robots can be roughly formulated as:*

- *All robots stay inside the field.*

- *All robots stay outside the opponent's penalty area.*

- *Each robot stays outside a* 3 m *radius around the ball or is within the own penalty area.*

- *If one robot is within the own penalty area, it is the only one.*

- *No opponent may be blocked by more than one robot.*

- *Each robot must block one of the four opponents by positioning in between it and the ball, unless that is not possible with respect to the other rules.*

- *One robot must watch the ball, i.e., position itself so that it can observe a line from the ball to the opponent second closest to the ball.*

- *Robots can form a defensive line in front of the own goal.*

- *Robots must observe a minimal distance to each other.*

The corresponding constraint system is very lengthy, thus we omit it here, however in conjunctive normal form, it consists of over 90 *distinct atoms, occurring over* 2000 *times in about* 220 *clauses. $Rprop(\Sigma_\wedge, \max)$ solves this problem in roughly* 30 ms, *while $Chaff+Rprop(\Sigma_\wedge, \max)$ runs until aborted after* 5 min.

The key to solving such problems efficiently is to propagate information back from the local search to the SAT-solver. This line of thought was established by Ganzinger et al. as DPLL(T) approach [53], where a Davis–Putnam–Logemann–Loveland algorithm [36] is combined with a solver for a theory T, which is able to solve incrementally, backtrack, and provide explanations for inconsistencies. However, to our knowledge, iSAT is currently the only SMT solver able to deal with continuous non-linear constraints which involve transcendental functions. We leave this problem open for future work. However, in the following sections we will consider SMT solving techniques when introducing additional solver features such as solution tracking and coordination with other robots' solvers.

9.4 Realtime Considerations

In realtime scenarios, the utility of obtaining a result to a constraint query is deteriorating with time. At some point $t + \Delta t_1$ the result to a query posted at time t becomes useless. Further, at $t + \Delta t_2$, a new query is posted, and the algorithm should ideally no longer be occupied with trying to solve the first query. However, in many cases, the new query is very similar to the previous, since the underlying problem changes rarely, only the sensory data involved is updated with small consecutive changes, yielding small changes in the constants of the constraint query.

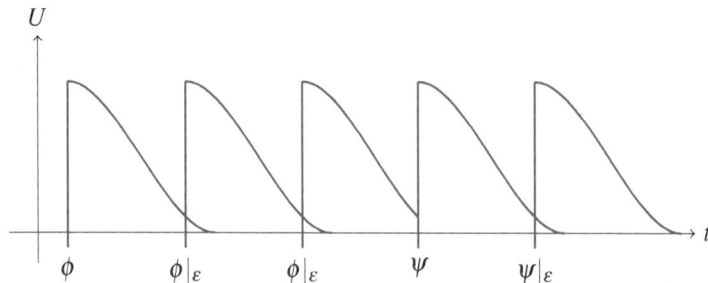

Figure 9.4: Utility of Solutions to Queries over Time – Three ε-equivalent formulae are queried followed by a new formula.

Figure 9.4 illustrates this effect on the utility of solutions to consecutively posted queries.

Definition 9.3. We say that two CSPs $(\phi, \vec{x}, C)$ and $(\phi', \vec{x'}, C')$ are ε-*equivalent* if and only if $\vec{x} = \vec{x'}$, $\phi = \phi'$ and C' can be obtained from C by replacing each constant k with another k' such that $|k - k'| \le \varepsilon$. By $\phi|_\varepsilon$ we denote any formula ε-equivalent to ϕ, i.e., the ε-neighbourhood of ϕ.

Example 9.2. *In the soccer domain, behaviours often run with a frequency of* 30 Hz, *as this is a common frequency of the most important sensor, the camera. Thus behaviours also post queries every* 33 ms. *Changes in the plan-tree, which potentially can cause the relevant constraint satisfaction problem to change significantly are much rarer. During RoboCup 2009, such changes were recorded in average every* 3.5 s *[157]. Let us assume a free-kick situation, in which one robot is positioning itself to receive a pass. In this case, a constraint describing the target position is queried in every iteration. After a few seconds, the referee signals the game to begin and the pass is played. Only then, the constraint problem changes significantly, as the robot now tries to obtain the ball as quickly as possible. Once it received the pass, the constraint problem changes again, this time to describe, say, a position to score a goal from.*

Although the constraint systems are non-linear, one can argue that if $\vec{p}$ is a solution to a constraint, then a solution to any ε-equivalent constraint problem can probably be found in the vicinity of $\vec{p}$. We refer to the ability of exploiting this relationship as *tracking*.

Tracking not only greatly reduces the time it takes to solve a query, but it also stabilises solutions over time. This is a highly important feature. Since the solutions to queries are typically used to obtain actuation commands, (e.g., when driving towards a constrained position), solutions should be as stable as possible to allow for smooth and efficient execution. Even when given noisy sensory data, solutions should not oscillate over time.

Dynamic constraint satisfaction problems, i.e., problems that change over time were discussed in depth by Brown and Miguel [16], who focused on the concept of local repair, where a solution in the vicinity of the old solution is sought. Additionally, they discussed the possibility of adding an oracle [172], which guides the search based on recorded information. However, they focussed on discrete constraint systems, while we deal mostly with continuous constraints. Nguyen and Yao [110] recently discussed genetic algorithms geared towards continuous dynamic constrained optimisation problems. They introduced a method of tracking feasible regions through a reference population whose members attract individuals outside of feasible regions. This approach is geared towards genetic algorithms and cannot be easily adapted to other approaches.

Various solvers can be enabled to track solutions through time. For local searches, which are initialised with a starting point, this can be done in a straight forward fashion. The first starting point is simply set to the previously solution. The local search will then ideally find a solution within only a few steps.

An SMT solver such as Chaff+Rprop($\Sigma_\wedge$, max) can also be extended to track solutions in one of the following ways:

- Cache both the propositional model that led to the last solution together with the last solution. In the next step, check if the cached interpretation is still a model. If so, use that model together with the cached solution to solve the non-linear problem. Intuitively, this approach has the advantage that the propositional interpretation tends to stay the same over time, hence the behaviour emerging from the constraint satisfaction problem can potentially be more coherent. It is however difficult, and in some cases impossible to match literals from one query to the next when the structure of the problem changes.

- Alternatively, one can cache the solution and simply omit the SAT-solving step altogether for the first run of the local search, and only employ SAT-solving once that fails. This avoids the problem of changed formulae structure, but does not guarantee stability of the propositional interpretation.

- Finally, one can cache the solution and, when confronted with a new query, firstly calculate an interpretation based on the solution, and use it to initialise a local search for a propositional model. Thereby, the problem of identifying literals from different queries is circumvented. The local search employed can then try to minimise the necessary changes to arrive at a model. A possible algorithm, named *Local Changes* was presented by Verfaillie and Schiex [173].

Thus, whether or not a SMT-based solver is used, only the solution needs to be stored to enable tracking. Since all variables of a constraint satisfaction problem can be uniquely identified, as they are either plan variables or correspond to agent fluents, we can even do better by caching variable-value tuples. This allows tracking even when some problem dimensions change. For each new dimension (i.e., variable) in a posed query, a random value can be attached to the cached solution, similarly to the case where no solution can be found based on the cached solution and a fresh starting point is chosen.

Since queries can be posed from any part of the plan-tree, in order to relate variables as much as possible, caching is done at the highest level with respect to the bindings provided by states and plantypes. The function traversing the plan-tree constructing a query presented in Section 8.7 already substitutes variables for their higher level counterparts if applicable. Therefore, caching can work straightforward using the variables of the CSP and their calculated values.

Extending the algorithm in Listing 9.1 correspondingly is trivial. Before the first run of the local search, instead of initialising the search with a random point, the cache is queried for a value for every variable of the problem. If no value can be found, i.e., a cache miss occurred, a random value is used. After a solution has been found, the corresponding values are stored in the cache. In the next section, we will discuss how the notion of a cache can be extended to provide a degree of coordination in a multi-agent context.

9.5 Coordination

In pALICA, coherent beliefs about the allocations within each plan were sufficient to achieve coherent execution. In the presence of constraint satisfaction problems, this is no longer the case. As multiple agents execute a plan, each agent individually solves a CSP periodically. Since the behaviour of an individual agent depends on the solution of this CSP, solutions should be equivalent or at least similar throughout the team such that the team executes the program in a coherent fashion. However, solutions can deviate within the team for several reasons:

Many Solutions – Typical continuous constrained problems have many solutions, often even infinitely many. If there is no metric which gives preference to a specific one, agents can arrive at completely different solutions. Objective functions can be added to the satisfaction problem, yielding a constrained optimisation problem, in order to circumvent this problem. We will discuss this problem class in more detail in Section 9.6. However, constraint optimisation can be much harder than constraint satisfaction. Therefore an alternative, less computational expensive way to guarantee coherence would be beneficial.

Symmetry – One specific source for the many solutions problem is symmetry. For instance, if two robots a and b in the robotic soccer scenario are supposed to block opponents c and d, without further restriction, a could come to the conclusion that it should block opponent c and b should block d, while b calculates that a blocks d and b blocks c. This is an instance of the task allocation problem, however this time integrated into a CSP. While again, objective functions can be used to break symmetry, other, less expensive symmetry breaking methods that achieve coherence are preferable.

Sensor Noise – Since sensory information are noisy, each robot has a slightly different view on the world, thus the set of possible solutions differs within the team. However, unless the beliefs of the agents diverge, the intersection of these sets is non-empty, thus the existence of a solution upon which the whole team can agree can be hypothesised. A coordination approach should select such a solution, if it exists.

Intersecting CSPs – In ALICA, each agent maintains its own constraint store, which depends on the plans the agent inhabits and the transitions it traversed. Moreover, agents may query different variables. Thus, the set of queried CSPs within the team, although connected by some variables may vary greatly in dimensionality and clauses. Should the intersection of the corresponding feasible regions be significantly smaller than the union, achieving coherence becomes difficult. Obviously, if the intersection is empty coherence cannot be established.

Before we discuss possible solutions for these issues, we establish some terms with respect to the relation of constraint satisfaction problems to each other.

Definition 9.4. Given two CSPs $c_1 = (\phi, X, C)$ and $c_2 = (\phi', X', C')$ we say that

- c_1 and c_2 are *independent* iff $X \cap X' = \emptyset$,

- c_1 and c_2 are *compatible* iff there is a solution η for c_1 and a solution η' for c_2 such that $(X \cap X')\eta = (X \cap X')\eta'$

- c_1 subsumes c_2, written $c_1 \succ c_2$ iff for all solutions η' for c_2 there is a solution η for c_1 such that $(X \cap X')\eta = (X \cap X')\eta'$.

- c_1 and c_2 are *equivalent* iff $c_1 \succ c_2$ and $c_2 \succ c_1$.

Trivially, all independent CSPs are equivalent, and all independent feasible CSPs are compatible. Moreover, infeasible CSPs are subsumed by any CSP, and if a feasible CSP is subsumed by another CSP, they are compatible.

Let $C = \{(\phi_1, X_1, C_1), \ldots, (\phi_n, X_n, C_n)\}$ be a set of CSPs considered by a team of agent. A simple way of dealing with this set and the arising problems, is to construct the intersection $\Phi = (\bigwedge \phi_i, \bigcup X_i, \bigcup C_i)$, solve it centrally and distribute the solution. Obviously, the intersection is subsumed by each individual CSP. This, however, introduces a bottleneck and — at least temporarily — a single point of failure. Moreover, in dynamic environments, it might be impossible to communicate the individual CSPs in time before new sensory information is available, and the global CSP is outdated before it can be solved.

Alternatively, similarly to the pALICA approach in task allocation, each individual agent can broadcast its solution periodically. Received solutions can then be integrated as variable-value tuples into the cache. Thereby, the cache accumulates solutions calculated by all robots involved. Depending on the CSPs and the belief bases involved, these can be tightly clustered or scattered within the solution space. Since the situation changes dynamically, an agent can never blindly trust a received solution, but should always check it against its current constraint store before usage, however, it is likely that a solution is nearby, assuming the CSPs do not vary greatly within the team. This suggests that received values can be used to derive initial points from which a local search can start to look for a solution.

Using each received solution as an individual starting point leads to poor performance in cases where these solutions are close to each other, since the same part of the search space would be investigated multiple times. On the other hand, merging all cached values into a single starting point is bound to overgeneralise in cases where the solutions are scattered across the problem landscape. Hence, solutions should be clustered using a dynamic number of clusters. Many initial points should be used when the solutions are scattered and hence the degree of conflict within the team is high, thereby allowing each agent to evaluate many or all solutions provided by the team and compare them locally. In cases where the solutions already converged to some small areas, fewer initial points are needed.

There are many possible clustering algorithms which can achieve this behaviour, such as hierarchical clustering based on the work of Ward [176], or sequential clustering with Bayesian filtering as proposed by Schubert and Sidenbladh [146]. In this case, the resulting cluster centres are used as starting points for a local search, hence the quality of the clustering result is less important than the time efficiency of the algorithm. Moreover, the results should be identical over the set of agents involved. Therefore, we employ a simple distance-based clustering, which may result in too many cluster centres. However, as this effect will only lead to a region being searched multiple times, it should not degrade the overall performance significantly.

We assume that the cache stores for each agent and for each variable the last received value, and the last computed value of the local agent. Listing 9.2 depicts a simple, but fast clustering in pseudocode. We use the following notational conventions: C denotes the cache and $C(a,x)$ denotes the cached value calculated by agent a for variable x. Finally, $C(a,x) = \bot$ denotes that no such value exists in the cache, i.e., no value for x has been received from a.

While the depicted algorithm cannot compete with established clustering approaches in terms of accuracy, it is fast and suffices to cluster points close to each other in the solution space, moreover, it can deal with partial solutions, i.e., solutions to CSPs, which do not include all variables queried. Such situations can arise when two agents solve intersecting CSPs, or one agent extends the CSP it considers with additional variables. The algorithm is not commutative, i.e., the sequence in which points are presented matters. Hence, the list of solution vectors, M, is uniquely ordered such that every agent is processing solutions in the same order.

The result is an ordered set of vectors in $(\mathbb{R} \cup \{\bot\})^n$, where $\bot$ indicates that no suitable value could be obtained from the cache. In order to obtain initial points for the local search, these values are replaced with random numbers. The set is sorted such that clusters with a higher number of supporters occur before clusters with a lower number. This results in a behaviour similar to a majority vote, such that each agent tends to prefer solutions calculated by many other agents. On the basis of this clustering we can extend the local search from Listing 9.1, which yields the algorithm in Listing 9.3.

This algorithm obtains the clustered results from the cache and iterates through them until the local search can obtain a solution. Should it run out of initial points it continues with random ones. The process is aborted once a timeout is reached, in which case solutions become irrelevant, and the CSP can be treated as if it was infeasible, according to the real-time considerations in Section 9.4. However, the algorithm performs at least one run of the local search for the first cached value and at least one run based on a random point, if no solution could be obtained so far. This guarantees that the most relevant (i.e., first) solution is tracked and that at least some exploration is performed in each cycle, until a solution is found. Thereby, the algorithm is allowed to violate the soft realtime constraint defined by the timeout in order to explore the search space.

```
InitialPoints (v⃗) {
// v⃗ is the ordered list of queried variables
Let r⃗ denote the size the domains of v⃗.
Let θ be a fixed threshold, i.e., 10⁻³.
R := ∅;

Let M be a list containing for all agents a ∈ 𝒜:
    c⃗ := (C(a, v₁), ..., C(a, v_dim(v⃗)));
Let M be uniquely ordered according to an ordering over 𝒜.

remove all (⊥, ..., ⊥) from M;

while (M ≠ ∅) {
    s⃗ := (⊥, ..., ⊥);
    k⃗ := (0, ..., 0);
    foreach (c⃗ ∈ M) {
        d := 0;
        t := 0;
        foreach (i ∈ [1, dim(v⃗)]) {
            if (cᵢ ≠ ⊥ ∧ sᵢ ≠ ⊥) {
                d := d + (cᵢ - sᵢ)²/rᵢ;
                t := t + 1;
            }
        }
        if (t = 0 ∨ d/t < θ) {
            foreach (i ∈ [1, dim(v⃗)]) {
                if (cᵢ ≠ ⊥) {
                    if (sᵢ = ⊥) {
                        kᵢ := 1;
                        sᵢ := cᵢ;
                    } else {
                        sᵢ := (kᵢsᵢ + cᵢ)/(kᵢ + 1);
                        kᵢ := kᵢ + 1;
                    }
                }
            }
            remove c⃗ from M;
        }
    }
    if ((∃sᵢ ∈ s⃗)sᵢ ≠ ⊥) R := R ∪ {(Σkᵢ, s⃗)};
}
Sort R descending by first tuple element;
return R;
}
```

Listing 9.2: Clustering of Cached Solutions

```
Solve(csp) {
    // parameters: s_init = 10^-3, inc = 1.2, dec = 0.5
    //  minV = 10^-12, maxV = 10^10
    begin := now;
    f := transform(csp);   // transform the problem into a function
    let p^min and p^max be the vectors of the lower (upper) bounds, respectively;
    d := number of dimensions of the csp;
    i := 1;
    M := InitialPoints (vars(csp));
    while( true ) {
        p := (⊥,...,⊥);
        if (i ≤ |M|) {
            p := M_i;
        }
        foreach (p_i ∈ p) {
            if (p_i = ⊥)
                p_i := random value ∈ [p_i^min, p_i^max];
        }
        j := 0;
        let s and h be vectors with d elements;
        set all s_k := s_init · (p_k^max − p_k^min);
        set all h_k := 0;
        while(++j < maxIterations ) {
            if (f(p) > 0) {
                insert p into cache;
                return p;
            }
            g := ∇f(p);
            for(k=0; k<d; k++) {
                if (g_k · h_k > 0) s_k := s_k · inc;
                if (g_k · h_k < 0) s_k := s_k · dec;
                s_k := min(max(s_k, minV), maxV);
                p_k := p_k + sgn(g_k) · s_k;
                p_k := min(max(p_k, p_k^min), p_k^max);
            }
            h := g;
        }
        if (time since begin > timeout) {
            if (i = 1) i := |M|+1; // do at least one free exploration
            else return FAILURE;
        }
        i := i+1;
    }
}
```

Listing 9.3: Local Search with Coordination through Caching

Due to the iteration over the clustered solutions, preference is given to solutions according to the total order over agents. This order determines the precedence among agents and could, in principle, be dynamically calculated according to the quality and relevance of their sensory information. However, it is most important that the order is known throughout the team. Here, we refrain from discussing how such dynamic orderings can be achieved and assume a simple, static ordering, e.g., based on unique ids.

This algorithm solves the problems of symmetric solutions, many solutions, and counteracts noise, as we will show in Section 10.3. Moreover, it utilises the distributed computing power of the team to search for solutions, effectively solving hard CSPs, which cannot be solved in one cycle, in a distributed manner. In this case, agents will explore the search space using random starting points, until one agent finds a feasible point, which is then distributed.

However, the algorithm does not address the problem of intersecting CSPs. While it can deal with CSPs that have different, but intersecting sets of variables, it does not guarantee convergence of the team when the feasible regions are different. This is a hard problem, that cannot be solved this easily (otherwise, the CSPs would be tractable). In other words, if a method existed that solves the intersecting CSP problem, which is more efficient than just constructing the intersection and solve it locally, than that method would be a more efficient solver in itself.

It is easy to see that no better approach can exist that does not sacrifice reactivity, given communication latencies, non-linearity, and incompleteness of the employed solver. Due to incompleteness, a solver cannot deduce infeasibility of a CSP. Basically, it can only test whether a given assignment constitutes a solution. Furthermore, an agent cannot know whether an interpretation is a solution to a CSP another agent considers as active without either knowing the CSP and testing the point locally or querying the other agent. Querying takes time due to latency and therefore risks that the problem changes in the meantime. If, on the other hand, the CSP is known by the querying agent, it could as well solve the intersection of the two CSPs. This is more efficient than considering the two CSPs separately, otherwise a better solver could be constructed by problem decomposition. Note that if one can assume that the communication latency is low with respect to the dynamics of the environment, messages can be used to implement distributed solving and optimisation techniques as discussed by Petcu [121].

Another possible approach to this type of coordination problem would detect a lingering conflict stemming from intersecting CSPs by monitoring the cache, and, should one occur, switch the decision protocol to another that sacrifices some reactivity in order to resolve the conflict. For instance, an agent can be appointed to construct the intersection, solve it, and broadcast the solution, similarly to the local leader protocol discussed in Chapter 6. Alternatively, each agent could assume explicit control of a part of the CSP and exchange its solutions with the team, similar to the DCOP approach discussed in [121]. We deem such a treatment of intersecting CSPs as future work.

9.6 Constraint Optimisation

In the previous sections, we focussed on constraint satisfaction problems, where the agents' behaviour is determined by the set of possible solutions to these CSPs. In many scenarios, however, one might not only want to describe a set of solutions, but also select the best among them according to some criteria. While this can still be represented as a constraint satisfaction problem by allowing for quantifiers, a formulation as a constrained optimisation problem is often much easier to solve.

Quantified CSP (QCSP) over Boolean domains are PSPACE complete [31]. To our knowledge, there is currently no algorithm able to tackle QCSPs over continuous domains effectively enough to be used by teams of agents acting in dynamic domains, especially in the presence of non-linear and transcendental functions. Recent work on solvers for QCSPs over continuous domains was done for instance by Goldsztejn et al. [60, 59].

Instead of following the QCSP line of thought, we allow for an optimisation criterion to be formulated together with a CSP, yielding a constrained optimisation problem (COP).

Definition 9.5. If $P = (\vec{x}, \phi, C)$ is a CSP, f a function mapping members of the domain of $\vec{x}$ to $\mathbb{R}$ then $P' = (P, f)$ is a COP. Any solution s to P is also a solution to P'. An *optimal* solution s to P' is a solution to P such that for all solutions s' to P, $f(s) \geq f(s')$. f is called the *objective* function.

In this work, we limit ourselves to the case of a single objective function, thus this definition excludes multi-objective optimisation.

Extending the local search from Listing 9.1 to handle optimisation problems is straight-forward. Essentially, instead terminating upon identifying a feasible point, the algorithm follows the gradient of the objective function inside feasible regions. The downside is that this does not yield a suitable termination criteria pure satisfaction problems have, and indeed, if the objective function is unbound within the search space, the algorithm cannot guarantee that it found a global optimum due to its incompleteness.

However, for many problems, an upper bound for the objective function is known in advance, or can be computed by interval propagation. Furthermore, for many scenarios finding the optimal solution is not necessary, but finding one *sufficiently* close to it is acceptable, especially when tight realtime constraints are given. We therefore assume that for any objective function f, there is a limit f_{max}, which denotes a sufficiency threshold, such that for any COP with objective function f a solution x such that $f(x) > f_{max}$ denotes a solution sufficiently close to the optimum that the search can be terminated. Of course, f_{max} may equal ∞, in which case the search will not terminate before a timeout is reached.

Example 9.3. *For a CSP, which constraints goal positions of a set of n robots, a suitable objective function minimises the sum of the squared distances from the robots' current positions to their goal positions:* $f(\vec{x}) = -\sum_{i=1}^{n} Dist(Goal(i), Pos(i))^2$.

This way, the robots minimise the distance they need to travel, allowing them to reach the described goal state swifter. Since the minimal distance between any two points is 0, the global optimum of $f(\vec{x})$ is 0, although this point is only feasible when the robots already reached feasible positions. Furthermore, due to sensor noise and accuracy limitations of the actuators, a robot cannot position itself arbitrarily precise. Therefore, a sufficiency limit of, e.g., $f_{max} = -0.1 \cdot n$ may be sensible, depending on the domain.

Optimisation also requires a slightly more complex treatment of exchanged or cached solutions, since solutions obtained from different initial points can now be compared with respect to their objective value. The exchange of solutions allows the team to distributively explore the search space, and obtain an optimal or near optimal solution quicker than a single agent can.

Thus, an agent should adapt the solution with the best objective value. However, to stabilise results over time and within the team, a certain preference should be given to solutions calculated earlier or by multiple other agents. In the case of satisfaction problems, this was achieved by the preference ordering over clustered solutions. This introduces two optimisation criteria, namely the given objective function, and coherence over time and within the team. Due to the dynamics of the domain, the two criteria can be in conflict.

We tackle this issue in the following way:

- Firstly, preference is given to the highest objective value, allowing for highly reactive behaviour.

- Secondly, the first run of the local solver, which is initialised by the cluster centre with the highest number of votes, is executed with a different parameter set, i.e., with a lower minimal step size and a higher number of iterations. This increases the chance of tracking an optimum.

- Thirdly, solutions stemming from the first run are given a slight preference, such that other solutions must be at least a certain threshold better than the one found during the first run. This hysteresis dampens oscillation.

- Finally, in the presence of an objective function, the unconstrained problem is optimised once and the resulting interpretation is used as an initial point for the constraint case. Thereby, the algorithm specifically searches for a feasible region near an optimum of the objective function.

These measures avoid the complications of solving a multi-objective optimisation problem which changes quickly over time and allow for quick reaction in case the objective landscape changes. In Section 10.3, we will evaluate this technique.

Listing 9.4 depicts the complete algorithm. The inner optimisation loop is shown in Listing 9.5. Note that the parameters for precise and less precise search are example values chosen ad hoc. The single optimisation run of the unconstrained problem is debatable, as it can constitute an improvement as well as an impairment (see [180] for a discussion on the underlying problem). However, it guarantees that in each iteration, time constraints permitting, a region with a high objective value is searched. We deem this beneficial to the overall problem.

While extending the SMT-based approach to optimisation is out of scope of this thesis, we like to mention some possibilities to deal with optimisation. Firstly, an SMT solver can iterate over all propositional models and compare the resulting solutions. This becomes quickly infeasible as the number of models increases. Secondly, an optimisation of the unconstrained objective function can be used to compute a propositional assignment as starting point for a local search iterating over the propositional neighbourhood. Finally, branch-and-bound techniques [32] can be used to narrow the propositional search space by iteratively asserting additional constraints whenever a solution is found.

9.7 Constraints and Task Allocation

The presence of CSPs and COPs in ALICA allows for powerful first-order decisions to be formulated in a concise way. Moreover, these can be solved efficiently. However, during plan execution agents are now confronted with two separate, but interleaving problems, namely finding an assignment to the queried variables and finding a task allocation mapping agents onto tasks. The solutions to both have a major impact on the resulting behaviour the agents exhibit. Typically, CSPs refer to dynamic properties of agents allocated to certain tasks, such as their positions within the environment, their remaining battery power, or the current configuration

```
Solve( satisfaction problem csp, objective function o, objective threshold o_t) {
  // parameters: significance threshold s_t = 10^{-22}
  begin := now;
  f := transform(csp);  // transform the problem into a function
  let p^{min} and p^{max} be the vectors of the lower (upper) bounds, respectively;
  d := number of dimensions of the csp;
  i := 1;
  M := InitialPoints (vars(csp));
  p_best := FAILURE;
  while( true ) {
      p := (⊥,...,⊥);
      if (i ≤ |M|) {
          p := M_i;
      }
      foreach( p_i ∈ p) {
          if (p_i = ⊥)
              p_i := random value ∈ [p_i^{min}, p_i^{max}];
      }
      if (time since begin < timeout ∧ i = |M| + 1) { //do one unconstrained optimisation
          p := Rprop(p,1,o,⊥);
      }
      p := Rprop(p,f,o,i=1);
      if (p ≠ FAILURE ∧ (p_best = FAILURE ∨ o(p) > o(p_best) + s_t) {
          p_best := p;
          if (o is constant ∨ o(p_best) > o_t) {
              insert p_best into cache;
              return p_best;
          }
      }
      if (time since begin > timeout) {
          if (i < |M| + 1) i := |M|+1; //do at least one free exploration
          else {
              insert p_best into cache;
              return p_best;
          }
      }
      i := i+1;
  }
}
```

Listing 9.4: Local Search with Optimisation and Coordination

of some of their joints. Hence, task allocation potentially influences the feasible regions of a CSP and has an effect on the optimal solution to a COP. This relationship is determined by the unfolding step (see Definition 8.8).

```
Rprop(p,c,o, precise ) {
    // arguments: p − the initial point, c − the transformed csp, o − the objective function,
    // precise − indicating which parameter set should be used
    // parameters: inc=1.2, dec=0.5, s_init = 10^{-3}, maxV = 10^{10}
    if ( precise ) {
        maxIterations := 110, minV := 10^{-15}
    } else {
        maxIterations := 60, minV := 10^{-11}
    }
    let p^{min} and p^{max} be the vectors of the lower (upper) bounds, respectively;
    d := number of dimensions of the csp;
    let s and h be vectors with d elements;
    set all s_k := s_init · (p_k^{max} − p_k^{min});
    set all h_k := 0;
    p_best := p;
    i := 0;
    while(++i < maxIterations ) {
        if (c(p) ≤ 0) {
            g := ∇c(p);
        }
        else {
            if (o is constant ) return p;
            g := ∇o(p);
            if (c(p_best) ≤ 0 ∨ o(p) > o(p_best))
                p_best = p;
        }
        for (k:=0; k<d; k++) {
            if (g_k · h_k > 0) s_k := s_k · inc;
            if (g_k · h_k < 0) s_k := s_k · dec;
            s_k := min(max(s_k, minV), maxV);
            p_k := p_k + sgn(g_k) · s_k;
            p_k := min(max(p_k, p_k^{min}), p_k^{max});
        }
        h := g;
    }
    if (c(p_best) ≤ 0) return FAILURE;
    return p_best;
}
```

Listing 9.5: Single Optimisation Run

Ideally, these two problems would be unified. The resulting problem statement would be to maximise an objective function which depends on the task of each agent as well as a set of continuous variables, such that a Boolean combination of constraints is satisfied.

A naive approach to combining the two problems would be: Iterate over all task allocations allowed by the cardinalities, and among the ones whose CSP is satisfied, take the one with the highest utility. However, this approach quickly becomes computationally infeasible. Moreover, if the global optimum is not found in the first iteration by the methods presented so far, one can expect a higher rate of task reallocations from this approach, degrading the performance of the team.

In principle, this problem can be treated as a mixed integer non-linear programming problem. Techniques tackling such problems have recently been improved by Nema et al. [107]. However, identifying a suitable solver for this kind of problem, which is able to perform well under tight time constraints (e.g., solve standard situation problems in less than 30ms) is future work. Empirical results presented by Nema [108] seem to indicate that these solvers are not suitable for such soft realtime scenarios.

It is possible to interleave the A^* search for a task allocation with constraint solving such that large portions of the search tree can potentially be pruned early. However, from another perspective, a unification of these two central problems has a major drawback. The task allocation algorithm presented in Section 5.12 is sound and complete. A combination with an incomplete algorithm, such as a local search for non-linear constraint satisfaction problems, can lead to an incomplete task allocation, which we deem highly undesirable.

The disjoint treatment of task allocation and constraint optimisation leads to an optimal solution whenever the best valid task allocation yields a feasible COP whose global optimum is at least as good as any global optimum of any COP yielded by any valid task allocation. Currently, this property can only be achieved by careful formulation of the runtime conditions and utility functions involved.

9.8 Summary

In this chapter, we introduced an anytime solver for a rich class of problems, namely Boolean combinations of non-linear constraints over continuous domains featuring rational and transcendental functions. We deem this problem class suitable for a large variety of problems occurring in multi-robotic domains. The solver is based on a local search combined with auto-differentiation, which provides a gradient at each considered point in linear time. We showed how solutions can be tracked over time using a cache and extended this caching method towards a coordination approach by integrating a clustering algorithm, which provides a majority vote over potential solutions. Afterwards, we discussed the relevance of constraint optimisation problems, and consequently extended our algorithm to this problem class as well. The solver is integrated into the reference implementation discussed in Chapter 7. Moreover, we discussed the potential of SMT solving techniques in this scenario based on a motivating benchmark test. Finally, we identified task allocation and CSP solving as two related problems which should be viewed from a unified perspective.

Part IV

Assessment

10 Evaluation

In the previous chapters, we presented the multi-robot programming language AL-ICA, starting with a basic propositional variant in Part II, which we then extended with constraints over continuous values in Part III. Here, we pick up the design goals introduced in Chapter 1 and present corresponding evaluation results.

Firstly, in Section 10.1, we present experiences in modelling team behaviours, which were gained by the Carpe Noctem team while using the presented approach. These experiences led to some design patterns which proved to be robust and reusable. Section 10.2 focuses on robustness given unreliable network conditions. We present simulation results in the robotic soccer domain with different degrees of packet loss and packet delay. We also show the effect of conflict detection and resolution under these circumstances. In Section 10.3, the robustness of our constraint solving and optimisation technique is assessed. We evaluate the noise over time and the coherence within a team of agents solving problems with different degrees of sensory noise. Section 10.4 shows a sketch of how to apply ALICA in an extraterrestrial exploration scenario. Moreover, we present a method to formulate dynamic formations with which multiple robots can explore an area in a coordinated fashion. Finally in Section 10.5, we employ ALICA in a scenario drawn from the rescue domain. ALICA is used to coordinate and control a team of fire brigades in order to fight multiple fires within a city. We compare the performance of ALICA with a set of established methods for the task assignment problem and assess the scalability of ALICA with respect to the number of participating agents.

10.1 Modelling in RoboCup

ALICA has been successfully used since 2009 by the Carpe Noctem Robotic Soccer Team of the University of Kassel[1]. This allowed developers to gain experience in implementing plans and strategies using ALICA. These experiences culminated in certain design patterns which proved to be useful, reusable, and robust against the features of the domain, namely noise, unreliable communication, and agents breaking down. They also illustrate the expressiveness of ALICA. In the

---

[1]   http://das-lab.net

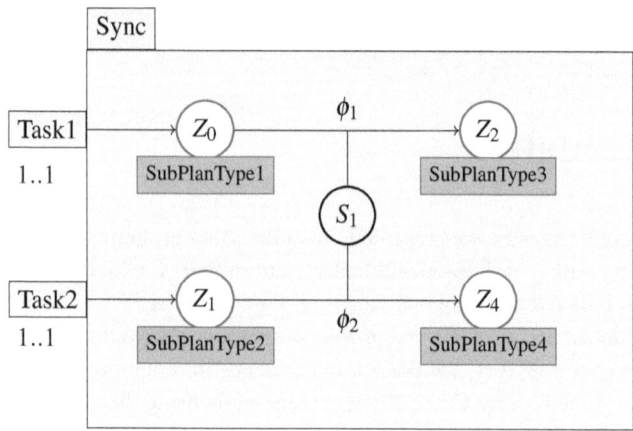

Figure 10.1: Strong Synchronisation in ALICA

following, we describe domain-specific formulae in a pseudo-formal manner, since domain-specific descriptions are not part of this work and a more formal presentation would not contribute to the given examples.

10.1.1 Strong and Weak Synchronisation

As mentioned earlier in Section 5.13, ALICA allows for strong synchronisation by employing explicit language elements, as well as weak synchronisation using conditions referring to the belief of the participating agents. Figure 10.1 depicts the usage of a synchronisation element to enforce strong synchronisation, such that agent will only move together from state Z_0 to Z_2 and Z_1 to Z_3, respectively. Agents will only move along these transitions if they have established mutual belief that ϕ_1 and ϕ_2 hold.

Strong synchronisation is rarely used in RoboCup due to the communication overhead. Weak synchronisation, on the other hand, is used frequently, for instance for when describing a pass. Figure 10.2 depicts the analogous pattern. Here, the agent executing *Task1* moves first along the transition, while the one executing *Task2* waits until it is notified that the first agent inhabits state Z_2. This kind of coordination requires only a single message. On the other hand, it does not truly synchronise the transition, since the second robot will always move after the first. This is acceptable in many scenarios. Weak synchronisation is also more susceptible to error, e.g., in cases where the second agent does not receive the plan-tree

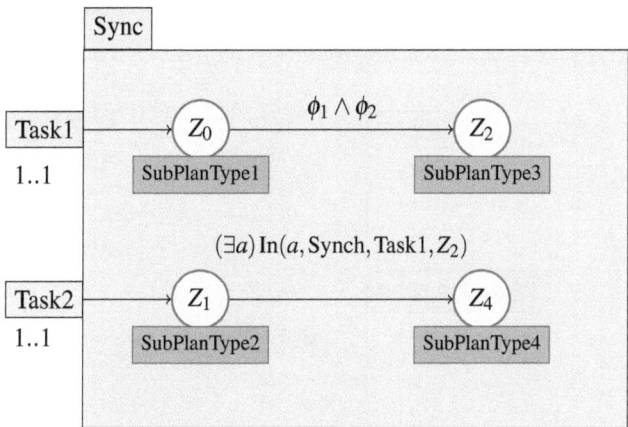

Figure 10.2: Weak Synchronisation in ALICA

message from the first. The periodic broadcast employed by ALICA alleviates this problem to a certain degree.

10.1.2 Finite State Machines and Dynamic Task Allocation

At the core of ALICA's modelling semantics are three diametrical concepts, namely finite state machines, dynamic task allocation, and constraint satisfaction and optimisation problems. The ability to combine these concepts arbitrarily is one of ALICA's strong points from a modelling perspective. Figure 10.3 shows a RoboCup strategy formulated using only dynamic reallocation. The corresponding utility function, $U(121Play)$, minimises the attacker's distance to the ball and the defender's distance to the own goal, thereby also prioritising the task *Defend* above the task *AttackSupport*. In contrast, the plan *Attack* depicted in Figure 10.4 is a pure finite state machine. This plan controls the robot's action given the perceived positions of the ball and the opponents.

These two instances show how both concepts, finite state machines and dynamic task allocation, are successfully used together within the same ALICA program. The two concepts can even be combined within the same plan, as shown in Figure 10.5. Here, two robots stay close to the dribbling robot and protect it against approaching defenders. The task allocation dynamically assigns agents to the three tasks depending on the situation, while transitions are used to switch the protecting robots between following and blocking.

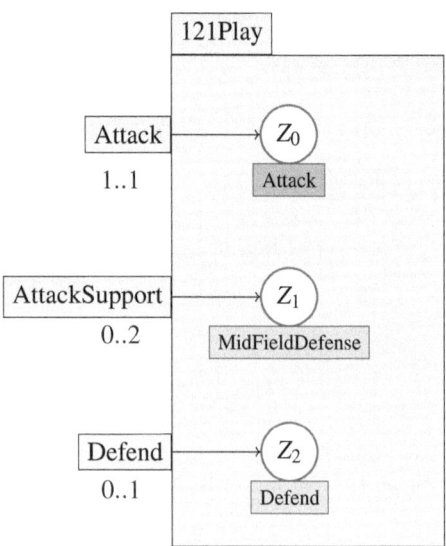

Figure 10.3: RoboCup Strategy 1-2-1

10.1.3 Select and Commit

The combination of finite state machines and dynamic allocation and decision making led to another design pattern, which we like to coin "Select and Commit". Here, the team executes a plantype and dynamically switches plans until a certain condition is met and the team moves to a state selected *depending on* the currently executed plantype, thereby committing to the dynamically made decision at the time the condition became true. Figure 10.6 illustrates this using an ALICA specification for a throw-in in robotic soccer.

The plan *ThrowIn* allocates one robot to the task *Keeper* and the rest to the task *FieldPlay*. While the keeper independently guards the goal, the field players either move towards their positions in state Z_1 or search for the ball in state Z_2 in case its location is not known. In this simplified example, the plantype *ThrowInPosition* contains four different plans, each meant for a different situation. The team selects one of them based on the ball's position, the behaviour of the opponent, and the current score of the game. As long as the game is in this positioning phase, the team can switch between these plans dynamically. The conditions ϕ_1 to ϕ_4 have the form

$$Situation(Start) \wedge ((\forall a, x, y) \operatorname{In}(a, ThrowIn, FieldPlay, Z_1) \leftrightarrow \operatorname{In}(a, p_i, x, y))$$

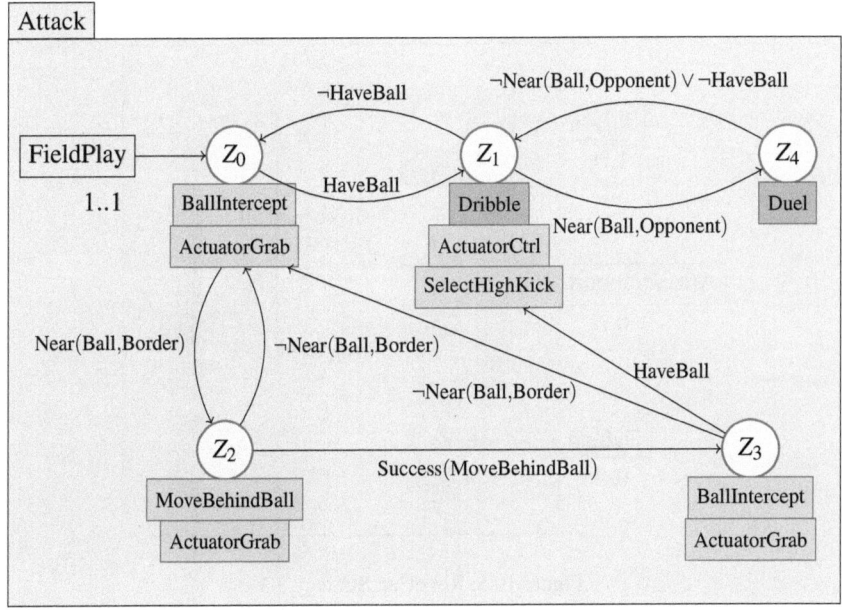

Figure 10.4: RoboCup Plan for Attacking Robot

where p_i refers to a specific plan in the plantype *ThrowInPosition*. Thus, upon reception of the start signal from the referee, the team will move to a subsequent state which is determined by the currently executed plan, provided that they believe to be in an agreement. If a conflict is known, the conflict is firstly resolved, and afterwards the field players move on to the execution phase. Note that this construct does not violate the locality principle, since the conditions appear on transitions. A similar construct within a precondition would violate plan locality.

The examples presented in this section illustrate the ease of modelling within ALICA as well as its expressive power. Due to the experience gathered by the RoboCup team, we see some design patterns emerging, similar to patterns in programming languages.

10.2 Unreliable Communication

We reported first evaluation results of pALICA without conflict detection and resolution in [157], focussing on the performance under poor network conditions. The

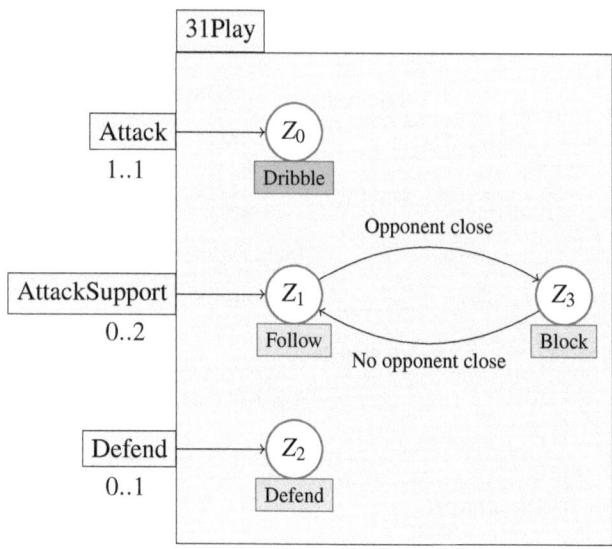

Figure 10.5: RoboCup Strategy 3-1

results showed stability under poor network conditions such as a uniformly distributed packet loss of up to 50% or packet delay of 600 ms. The simulated results for packet delay were confirmed under real-world conditions during the RoboCup World Championship 2009 [157].

One of the major changes within the ALICA framework since then were the addition of explicit conflict detection and resolution. Conflict detection and resolution in ALICA was first discussed in [158] and used since the beginning of 2011 within the Carpe Noctem RoboCup team. We could show that conflicts due to static error, i.e., inconsistent configurations can be reliably detected and solved by the resolution scheme presented in Chapter 6. Furthermore, the approach is also able to compensate conflicts due to systematic sensor errors, such as robots underestimating small distances to perceived objects.

In the following, we present an evaluation of ALICA's performance in unreliable network conditions with and without conflict resolution. In each experiment, we simulate four robots playing soccer for five minutes under different network conditions and measure the time from an event triggering a dynamic reallocation until the team achieves a coherent view on the task allocation. We refer to this time as *time to coordinate (TTC)*. This measure is similar to ATA (Average time to agreement) used by Kaminka and Tambe [82]. However, ATA is measured in

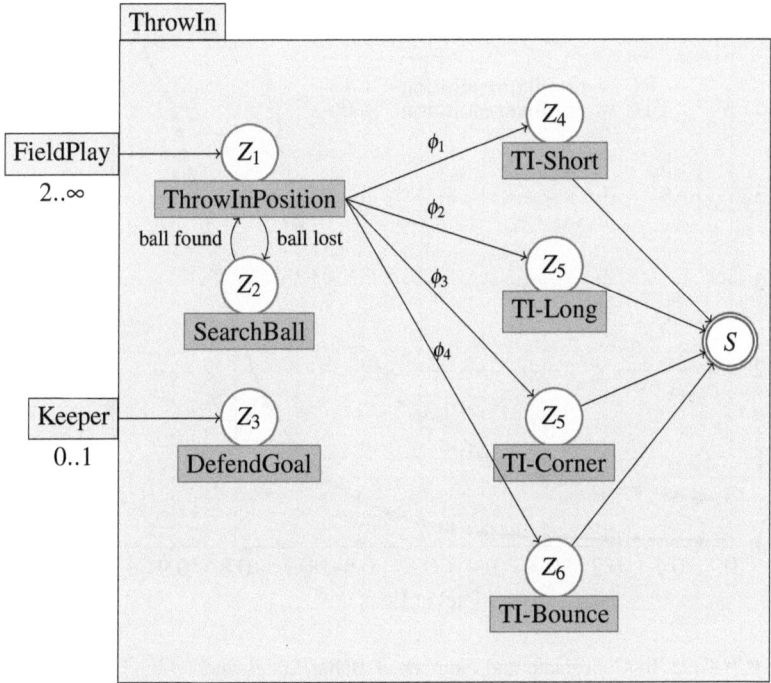

Figure 10.6: RoboCup Plan for Throw-In Situations

abstract time units called "ticks" introduced by the respective simulation, which also synchronises the deliberation cycles of the agents involved. TTC is measured in seconds. Furthermore, our simulation does not artificially synchronise actions or sensory information of the agents involved.

Additionally, we measured the average number of different belief states present within the team, called *Belief Count (BC)*, which can range from one (total agreement) to four (all agents have a different view). This measure indicates the amount of disagreement in the team. The more dynamic an environment is, the higher the average belief count should be, as agents cannot agree faster than they can make decisions and further, sensory information about the dynamically changing elements of the environment needs to be communicated to agents that cannot observe them directly. In robotic soccer, such elements include the ball, the opponents, and the positions of team members.

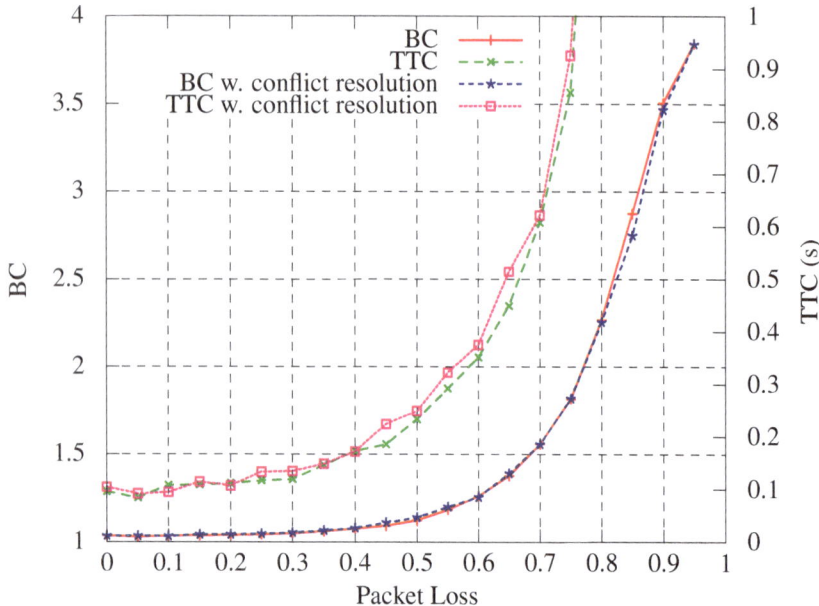

Figure 10.7: Time To Coordinate and Number of Belief States under Simulated Package Loss.

In the following experiments, the deliberation frequency of each agent is set to 30 Hz, the maximal communication frequency (f_{max}) to 15 Hz and the minimal communication frequency (f_{min}) to 5 Hz. For experiments with conflict resolution, the number of subsequent cycles triggering it is set to 4, and the authoritative time interval is bounded between $t_{min} = 0.8$ s and $t_{max} = 5$ s. In order to introduce further dynamics to the experiment, the simulator repositions the ball randomly every five seconds.

In the first experiment, the simulator emulates perfect sensors, i.e., no noise or errors were added to the observations. Figure 10.7 depicts the results under package loss and Figure 10.8 shows the results for artificially created packet delay. Each data point represents 45 min simulation.

We can draw the following conclusions from Figure 10.7: Firstly, we see that without conflict resolution under perfect network conditions, the TTC is about 100 ms with a BC of 1.04. Hence, on average, the robots achieve an agreement after one communication cycle and three deliberation cycles. Due to lack of per-

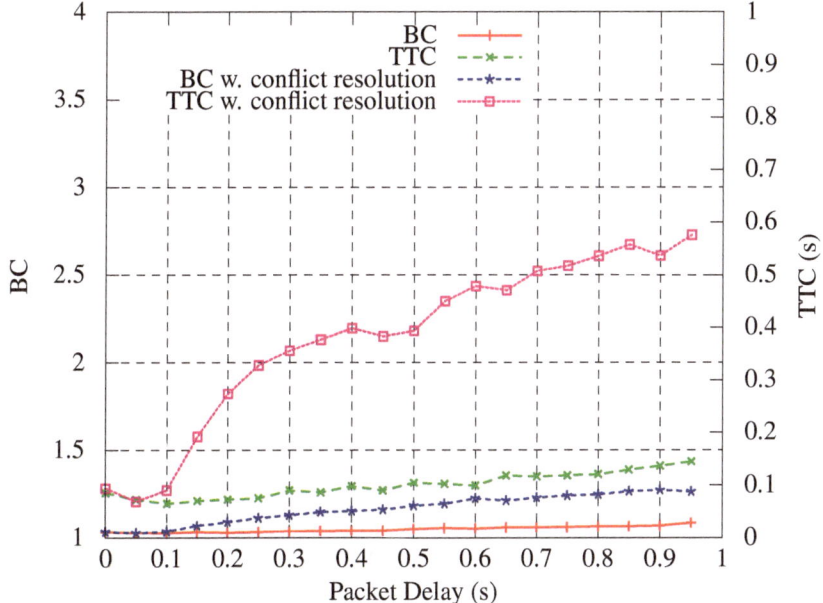

Figure 10.8: Time To Coordinate and Number of Belief States under Simulated Package Delay.

sistent conflicts in this setting, task allocation with conflict resolution shows the same performance. This is comparable to the ATA obtained by Kaminka and Tambe [82], 3.65 ticks, although in their experiments, team members were able to communicate in every deliberation cycle.

Secondly, the BC using task allocation with and without conflict resolution increases very slowly until a packet loss of 50% is reached, at which point it starts to grow exponentially. The same holds for the TTC value. Note that the higher packet loss, the more likely it is that an agent incorrectly assumes that others broke down due to lack of message reception (see Section 7.5.2). This effect appears to become dominant in this case at a packet loss ratio of about 60% to 70%.

In contrast to packet loss, introducing packet delay into the system has a completely different effect, as shown by Figure 10.8. Without conflict resolution, the team is much more resilient to packet delay than to packet loss. This is to be expected, as the robots mostly base their decisions on perfect local data provided by the simulator and thus arrive at the same conclusions. Therefore, the indicators only increase slightly with packet delay. However, when conflict resolution is used, packet delay has a significant impact on the time to coordinate. There are two reasons which in combination cause this effect:

- Packet delay introduces cycles in the task allocation history, since agents receive messages referring to past events and subsequently revert the corresponding update by dynamic reallocation. Therefore, packet delay triggers the conflict resolution protocol. This effect could be countered by taking advantage of synchronised clocks (i.e., by using the NTP protocol [101]) and discarding or modifying messages older than a certain age, thus improving performance in terms of team coherence.

- Conflict resolution relies heavily on communication. During authoritative mode, the task allocation problem is solved by a single robot which broadcasts its results. The delay of these authoritative messages has a high impact on the team's coherence. Every time the elected leader makes a decision, it is in disagreement with the rest of the team until the team receives and reacts on the respective message, a process that takes about 50 ms in addition to packet delay. In normal mode, agents can simultaneously react to the environment.

For the next experiment, we modify the simulator to add systematic error to the perceived ball position. Each robot perceives the ball 30 cm closer than it actually is. This systematic error roughly emulates real conditions, where the image processing overcompensates for poor lighting conditions. Under conditions with very little ambient or diffuse light and a high amount of direct light, the ball's lower half appears black and is no longer recognisable as part of the ball, which leads to an overestimation of the distance to the ball. In practice, this effect is often countered by heuristics, which in turn overcompensate the effect, making the ball appear closer than it actually is.

Figure 10.9 shows the resulting coordination under packet loss and Figure 10.10 shows the same for packet delay. The data clearly shows that conflict resolution is advantageous in cases where the beliefs within the team diverge, e.g., due to systematic sensory errors. Under perfect network conditions, the average number of belief states within the team is 1.2, almost 20% worse than in the previous scenario without systematic errors. Using conflict resolution, the BC is reduced to 1.08. The difference between the two coordination schemes is even more apparent in the TTC, where conflict resolution achieves agreement under perfect network conditions in 150 ms in average, while conflicts persist for 450 ms otherwise. Conflict resolution cannot achieve results as good as when perfect sensory data is available,

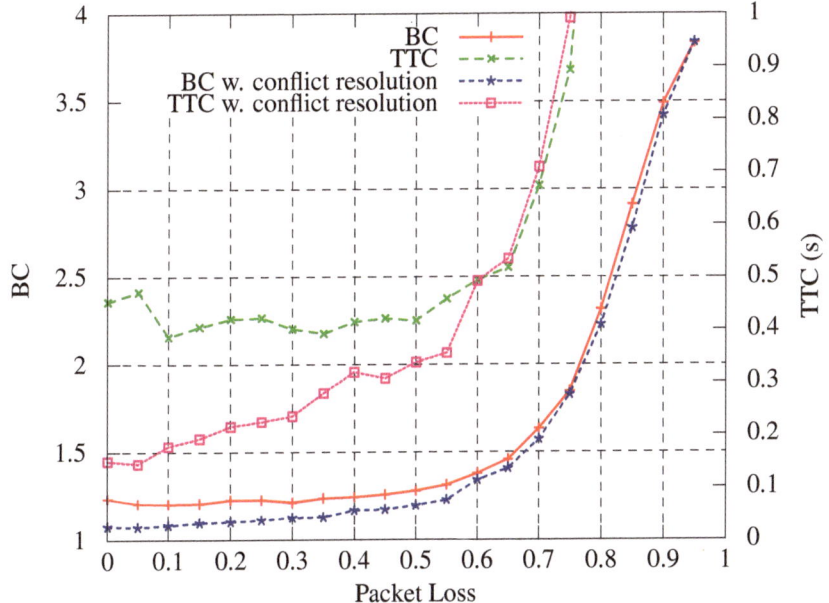

Figure 10.9: Coordination under Packet Loss with Systematic Errors

since conflicts first need to be detected and then communicated about before they are resolved.

Conflict resolution improves performance until a packet loss of 60%, at which point lost authoritative messages start to cause more conflicts than can be resolved. A similar trend can be observed for packet delay (Figure 10.10), where conflict resolution improves coordination up until at 300ms delay, at which point, the team starts to perform better without conflict resolution, due to its reliance on communication.

Interestingly, the TTC without conflict resolution actually improves with packet delay in this experiment. This is the result of a fairly complex interrelation between sensor fusion and packet delay. Packet delay increases the distance between the local observations and remote observations with respect to moving objects. At some point, the different observations are not merged any longer and are considered as two different objects by the sensor fusion algorithm. A dribbling robot is therefore perceived by its team members as an opponent dribbling, since they do match the remote localisation information with the local obstacle information. Since there

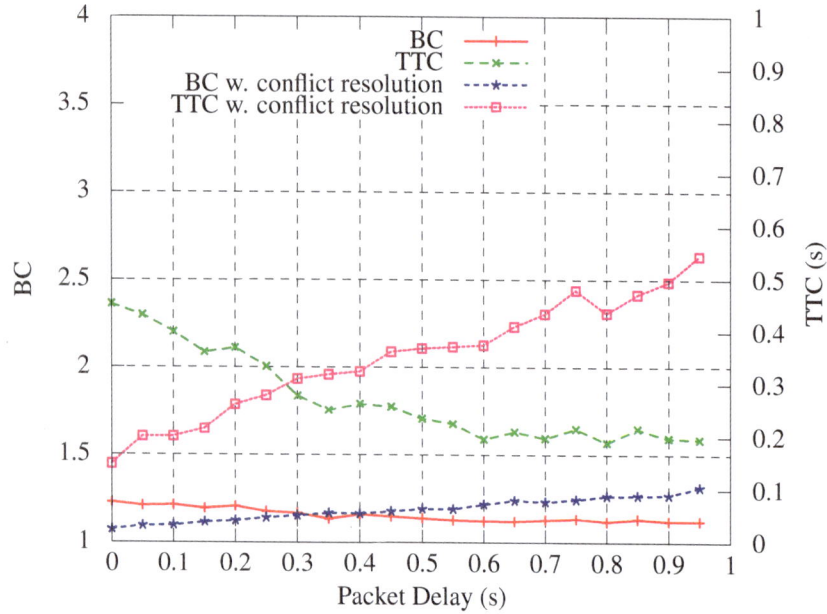

Figure 10.10: Coordination under Packet Delay with Systematic Errors

is exactly one ball in the game, the robots have to select one hypothesis. The precise calculations are based on the Dempster-Shafer theory of evidence [149] (see [131] for details regarding this scenario). In this situation, they favour the observation of the robot closest to the ball, i.e., the dribbling robot, since it has the highest confidence. Thus, other robots will rely on the delayed information sent by the dribbling robot instead of their own sensory information, provided that the distance to the ball, the delay, and the velocity of the dribbling robot are large enough. Thus, packet delay partially increases the level of coherence among the belief bases in the team.

We do not present results for scenarios with sensory noise. In our experiments, sensory noise caused the performance to degrade slightly, which is to be expected. Apart from that, the overall behaviour of the TTC and the BC resemble the case with perfect data.

In summary, with disagreement periods of 100 ms under perfect conditions, 230 to 250 ms under 50% packet loss, and 110 to 270 ms under 200 ms delay, we can state that the agents not only react quickly to changes in the environment but also achieve agreement about the task allocation quickly and hence act and cooperate according to the modelled plans. Moreover, systematic sensor error can be effectively countered by conflict resolution, given the network is not degraded too much.

In realistic scenarios, network quality cannot be regarded as constant as it was in these experiments. Instead, the network is subject to bursts of errors, which affect packet loss and delay for a short period of time. Under these circumstances, the performance of the team should quickly degrade in the same fashion as in the presented experiments if the burst is long enough. Short bursts are compensated for, since each agent computes team-level decisions in which it is involved individually and maintains its beliefs about them until notified of conflicting information or until it deems other robots to be out of order due to lack of messages for a longer period of time.

The degree of achievable coherence depends on the dynamic of the domain, the reliability of the network, and the coherence of the sensor information. In settings where all three factors are extremely demanding, every coordination approach will eventually fail. However, one can tackle extreme scenarios using additional means of coordination, such as action recognition or using the environment to communicate. Integrating action recognition along the lines of the work of Huber and Durfee [75] is future work.

10.3 Constraint Solving and Optimisation

One of the main features that distinguishes ALICA from other approaches to multi-agent coordination is the integration of constraint satisfaction and optimisation problems. In Definition 9.1, we proposed a problem class that suits a variety of robotic and multi-robotic problems, namely non-linear continuous constraint satisfaction problems (CNLCSP). Subsequently, we described a solver that can tackle this problem class in real-time scenarios and coordinate the results within the team. The solver as well as the problem class are meant to be exchangeable and appropriate interfaces are provided, so that for different domains, different problem classes can be used to describe team behaviour. However, the described algorithm is integrated into the ALICA reference implementation. In [155], we presented a detailed evaluation of the performance of this and other solvers using problems drawn from robotic domains. The solver discussed in Chapter 9 showed the best performance in these experiments, and was able to solve most of the benchmark problems in a time acceptable for near realtime reasoning.

The following series of experiments evaluates the stability against noise as well as the degree of coordination achieved by our distributed constraint satisfaction and optimisation approach. In each experiment, a team of agents solves a satisfaction or optimisation problem similar to those occurring in robotic scenarios. Each participating agent is individually affected by Gaussian noise. Similarly to real-world scenarios, each agent solves the problem in each deliberation cycle. After 500 iterations, the agents terminate.

We measure the standard deviation of each agent's solution over time and the standard deviation within the team at any point in time.

10.3.1 The Ring Problem

Firstly, we evaluate our approach on a simple constrained quadratic optimisation problem:

Let c be a fixed random point in $[-10^4, 10^4]^2$, find $p \in [-10^4, 10^4]^2$ such that

$$o(p) = 4 \cdot 10^4 - |c - p|$$

is maximised subject to

$$|c - p| \geq 2 \cdot 10^3$$

The set of optimal solutions is a circle with a radius of $2 \cdot 10^3$ around c. All points outside that circle are solutions. Note that the constants in this and the following objective functions are used to guarantee that the functions map to values larger than 1. This is due to an implementation detail in the solver, which simplifies the necessary computations by a small margin and has no other consequence. In the experiment, each participant sensed the point p under isotropic Gaussian noise with deviation σ_{in}.

Figure 10.11 depicts the resulting average standard deviation per agent over time. Note that in cases with one or two agents, this output noise increases drastically at high noise levels. At high noise levels, one or two agents are not able to reliably track previous solutions. However, with an increasing number of participating agents, the noise levels decrease drastically. This effect is most apparent when adding the third agent and afterwards shows diminishing returns. This behaviour is as expected, the exchange of solutions used as starting points for the local search and the majority vote over solutions stabilises the result as soon as a majority can exist. Increasing the number of available votes has less and less impact on the result.

Figure 10.12 shows average standard deviation within the team at any point in time, it is a measure for the team's level of coherence. The higher this noise

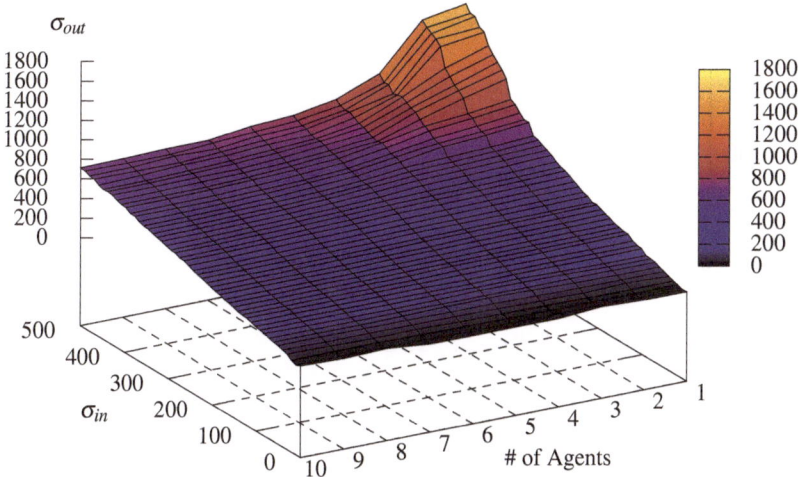

Figure 10.11: Resulting Noise Levels over Time in the Ring Problem

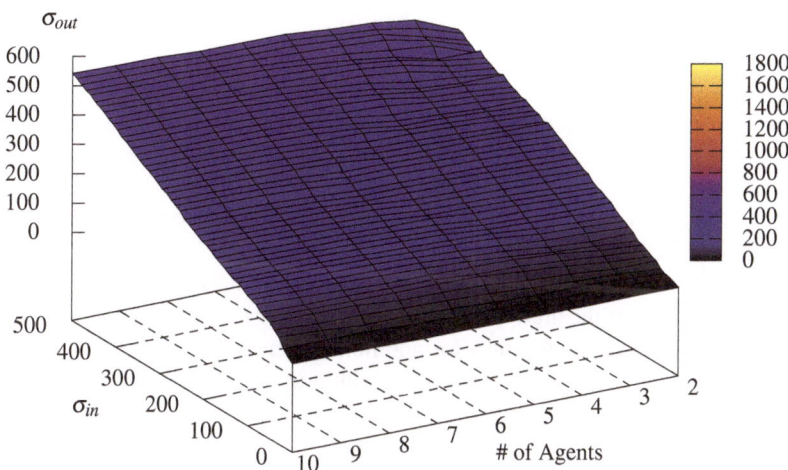

Figure 10.12: Resulting Noise Levels within the Team in the Ring Problem

measurement, the less coherent the team can act. Comparison between the two measurements reveals that the noise within the team is consistently lower than the noise over time. Therefore, even though the team as a whole is still susceptible to

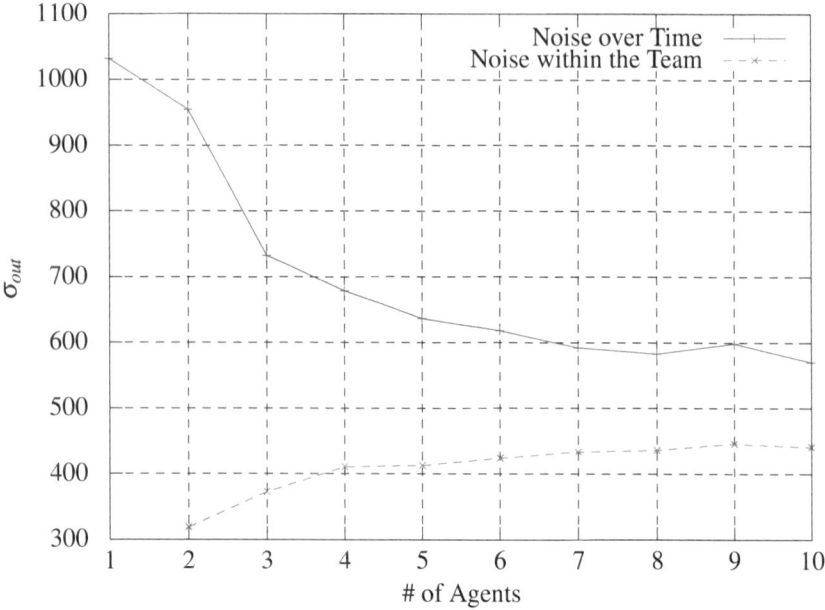

Figure 10.13: Resulting Noise Levels for $\sigma_{in} = 400$ in the Ring Problem

the sensory noise, it reacts coherently to it. Indeed, the noise within the team is only slightly higher than the input noise.

Figure 10.13 shows a 2D view on the two curves at an input noise level of $\sigma_{in} = 400$. It depicts the drastic reduction in noise over time when adding the third agent and the diminishing return for additional agents more clearly. Moreover, we only see a slight increase in the noise within the team as agents are added to it. This increase reflects the distribution of solutions due to the distribution of the input point c.

10.3.2 Blockers

As a second experiment, we use a simplified problem from the robotic soccer domain. Given the positions of four opponent robots and of the ball, the task is to find three positions on the field, which are at least two meters from the ball, at least one meter away from each other, and 0.7 m from away from one opponent along

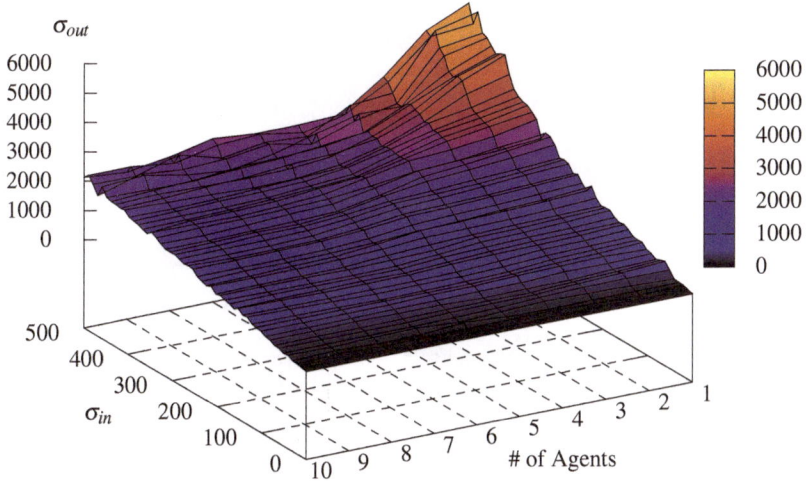

Figure 10.14: Resulting Noise Levels over Time in the Blocker Problem

the vector towards the ball. This is a disjunctive satisfaction problem without an objective function.

Formally, find $p_1, p_2, p_3 \in [-10^4\,\text{mm}, 10^4\,\text{mm}]^2$, given b, a fixed random point in $[-10^3\,\text{mm}, 10^3\,\text{mm}]^2$, and o_1, o_2, o_3, o_4 uniformly distributed fixed points between $2.5\,\text{m}$ and $8\,\text{m}$ away from b. The solution must satisfy:

$$((\forall i)|p_i - b| \geq 2\,\text{m}) \wedge \tag{10.1}$$

$$|p_1 - p_2| \geq 1\,\text{m} \wedge |p_1 - p_3| \geq 1\,\text{m} \wedge |p_2 - p_3| \geq 1\,\text{m} \wedge \tag{10.2}$$

$$\left((\forall i)(\exists j) \left| p_i - o_j + \frac{0.7\,\text{m} \cdot (b - o_j)}{|b - o_j|} \right| < 0.01\,\text{m} \right) \tag{10.3}$$

The quantifications in the last line over the indices i and j are used to abbreviate the corresponding conjunction and disjunction, respectively. Thus, the solution is constrained to be any combination of points in front of the four opponents with a precision of $1\,\text{cm}$. Typically, one would add an objective function to this scenario in order to minimise the distance each robot has to travel to arrive at the target positions. Here, we omit this to maintain a pure satisfaction problem and to avoid the noise inhibition such an objective function would entail. All observations, the ball as well as the opponents, are subject to isotropic Gaussian noise with deviation σ_{in}, measured in mm.

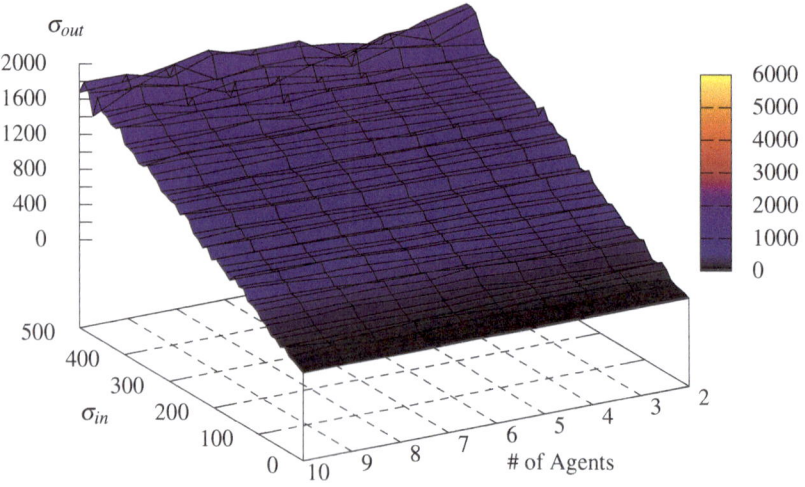

Figure 10.15: Resulting Noise Levels within the Team in the Blocker Problem

Figure 10.14 shows the resulting noise over time, which features the same characteristics as the simpler conjunctive optimisation problem in Section 10.3.1. However, the resulting noise level is much higher. This is to be expected since instead of a single noisy point in the equation, there are now five and each potential target position depends on two noisy points. Moreover, whenever the agents lose track of the solution, they may switch to a completely different one, i.e., a different case in the disjunction, thus inducing a large variance. In some instances, this is even required to maintain a solution, due to the second constrain requiring a minimal distance between the target positions. In some cases, the CSP can even become infeasible, namely, if the opponents appear too close to each other or to the ball. This happened in 0.042% of the cases, i.e., once every 45 s.

Similarly, Figure 10.15 shows the same trend as in the previous example. The noise within the team is consistently lower than the noise over time, thus the robots are coordinated even when they lose track of a solution. Figure 10.16 is a 2D projection of the noise over time with one to five agents. It shows again the sudden drop in noise once three agents participate in this problem and the diminishing return for every additional agent.

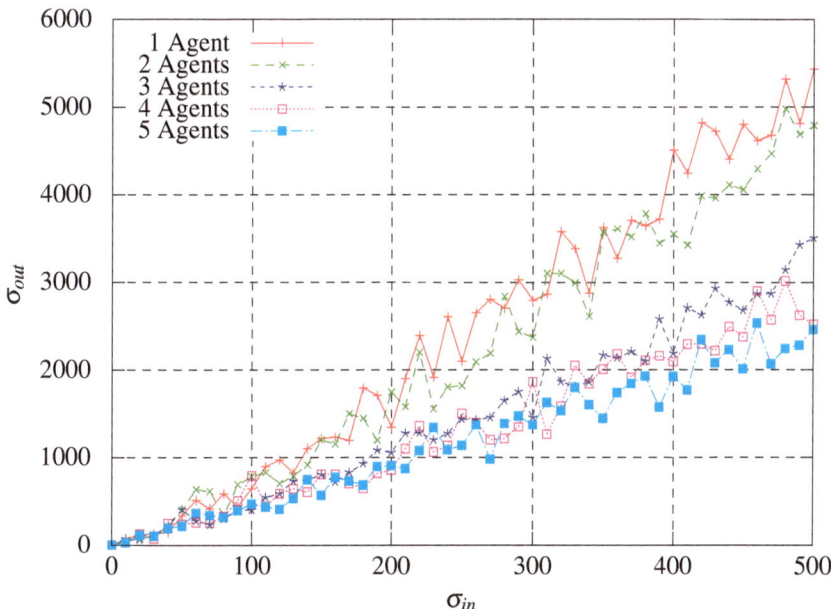

Figure 10.16: Resulting Noise Levels over Time in the Blocker Problem

10.3.3 Inverse Kinematics

As a last experiment, we use an inverse kinematics problem. Here, the goal is to position the end-effector of a Jaco™ robotic arm by Kinova. The arm features six degrees of freedom. The position of the endeffector corresponds to the point p in homogeneous coordinates:

$$p = M_1 \cdot M_2 \cdot M_3 \cdot M_4 \cdot M_5 \cdot M_6 \cdot (0,0,0,1)^T$$

where M_i is a matrix representing the transformation by the i-th joint of the arm. Each transformation depends on one degree of freedom. The orientation of the end-effector is defined by the transformation matrix $M = M_1 \cdot M_2 \cdot M_3 \cdot M_4 \cdot M_5 \cdot M_6$ with elements m_{ij}. Given the target point g and target rotation matrix R with elements r_{ij}, the objective function is:

$$U = 1000 - |g - p| - \sum_i \sum_j (m_{ij} - r_{ij})^2$$

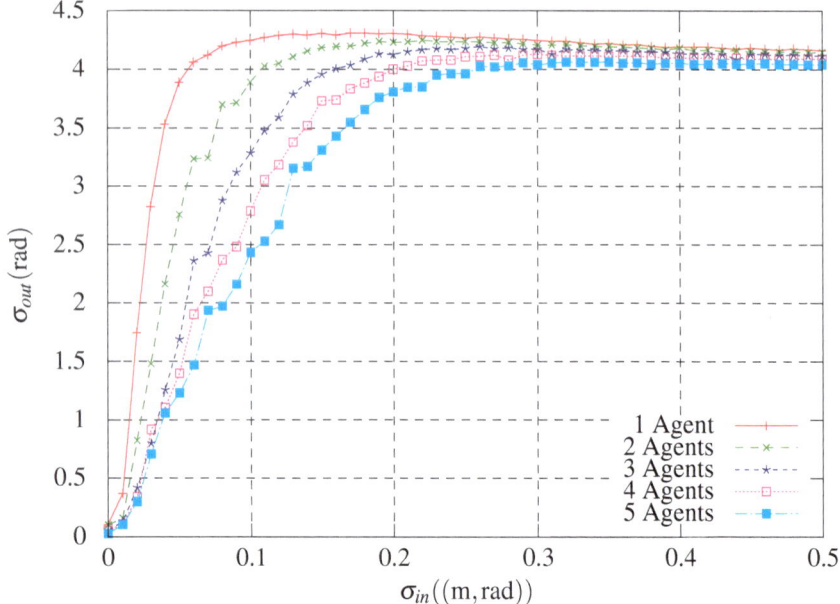

Figure 10.17: Resulting Noise Levels over Time in the Inverse Kinematics Problem

we do not add any constraints, thus evaluate a completely unconstrained optimisation problem. In the experiment, each individual dimension of both the target point and the target orientation is subject to Gaussian noise σ_{in}.

Figure 10.17 shows the resulting noise level over time with different numbers of agents. Note that the arm has a maximal range of about 1.5 m, so an input noise of 0.1 is already quite significant compared to the arm's operating range. Therefore, the resulting high noise levels are not surprising. However, we can see the decrease in output noise due to multiple agents solving the problem. Moreover, in line with the previous experiments, the diminishing return is clearly visible.

For comparison, Figure 10.18 depicts the noise levels within the team versus the noise level over time for the case of five agents. While both measurements react similar to input noise, noise within the team is lower than noise over time. Therefore, the team also coordinates in this unconstrained scenario.

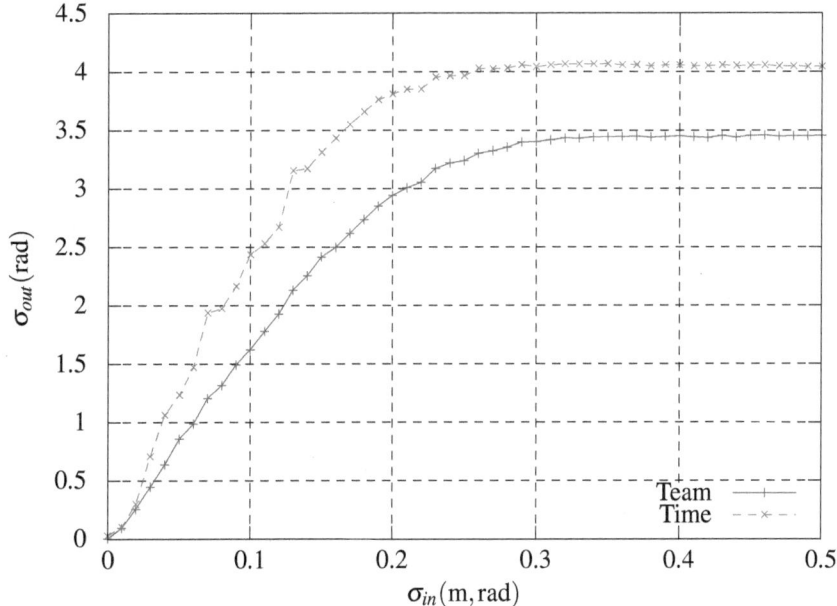

Figure 10.18: Resulting Noise Levels for 5 Agents in the Inverse Kinematics Problem

10.3.4 Summary

Summarising, we can state that the proposed algorithm coordinates the team, as evident by the experiments, in pure satisfaction problems, pure optimisation problems, and mixed problems. Individual decisions are still susceptible to noise to a degree that depends on the specific problem. Due to the trade-off between reactivity and robustness against noise, it seems unlikely that better results can be achieved without sacrificing reactivity to a certain degree. In all experiments, each agent solved the given problem with respect to its local, noisy sensory data. In other words, this approach focuses on reactivity. Note that this algorithm can trivially be combined with a smoothing or sensor fusion preprocessing step. We deem such a combination mandatory for all real-world scenarios.

10.4 Case Study: Exploration

In Section 1.3.2, we discussed extraterrestrial exploration as a possible target scenario for ALICA. Since such scenarios typically require highly specialised equipment for equally specific tasks, one can expect that any robotic team would be heterogeneous. That is, the team consists of different robots, each with a set of different abilities and actuators.

Representation and reasoning about heterogeneous teams is straight-forward in ALICA, since it allows the definition of roles based on required capabilities, and then matches these roles to tasks within plans that ought to be executed.

Consider the following lunar scenario: A small team of robots is tasked with finding and retrieving components of a communication station. The components are distributed in an unknown environment. The team consists of four small robots, equipped with various sensors for exploration, and two larger robots, which are equipped with arms, so they can grab and carry the components once found.

To distinguish between the two kinds of robots, one can introduce the following capabilities

- *Speed*: $\{\top, \bot\}$
- *CanGrab*: $\{\top, \bot\}$
- *CanCarry*: $\{\top, \bot\}$

based on these capabilities, one then defines roles within the team:

$$\mathcal{R} = \{Scout, Transporter\}$$

The role *Scout* requires robots to be fast, while the role *Transporter* requires robots to grab and carry objects. Role allocation will then match the smaller robots onto the role *Scout* and the larger robots onto the role *Transporter*. Note that this example is slightly simplistic, it is easy to envision a more heterogeneous team composition, where some robots are able to grab objects, but cannot carry them over longer distances. In this case, the role *Transporter* could be split into *Transporter* and *Loader*.

Figure 10.19 depicts how the problem can be represented as an ALICA plan on the highest level. Scouting robots would take on the task *Scouting*, while transporters would take on the task *Retrieving*. Both decisions are made based on preferences between roles and tasks, which encode whether or not a certain role can take on a certain task and how well it can potentially perform.

Scouts will execute the plantype *Search* until all components are found and then return to base. Transporters will idle in state Z_2 until a component is discovered, in

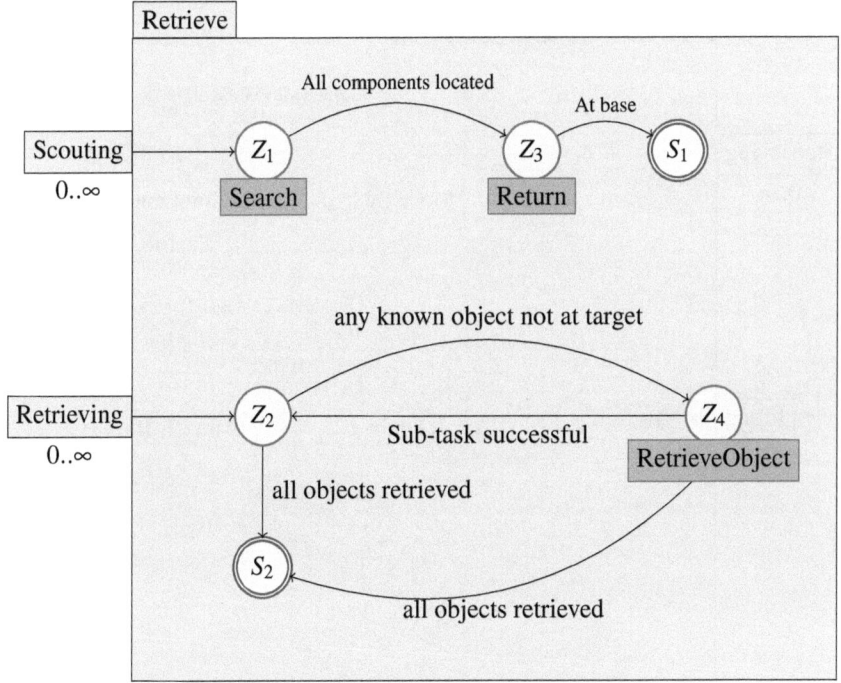

Figure 10.19: Plan for Search & Retrieval

which case they start to execute *RetrieveObject*. Whenever a task in *RetrieveObject* is completed, the corresponding robot will go back to the idling state Z_2. Once all objects are retrieved, all transporters successfully terminate this task.

The two plantypes *RetrieveObject* and *Search* are interesting examples illustrating the capabilities of ALICA further. We will firstly discuss *RetrieveObject*. As indicated by the cardinalities of the plan *Retrieve*, this plantype should work with any number of robots and objects.

10.4.1 Retrieving

In the remainder we assume the presence of the following domain predicates:

- *Distance*(a, b, d) – d is the distance between positions a and b.

- *ComponentsLocated*(n) – n is the number of located components, that are not yet transported.

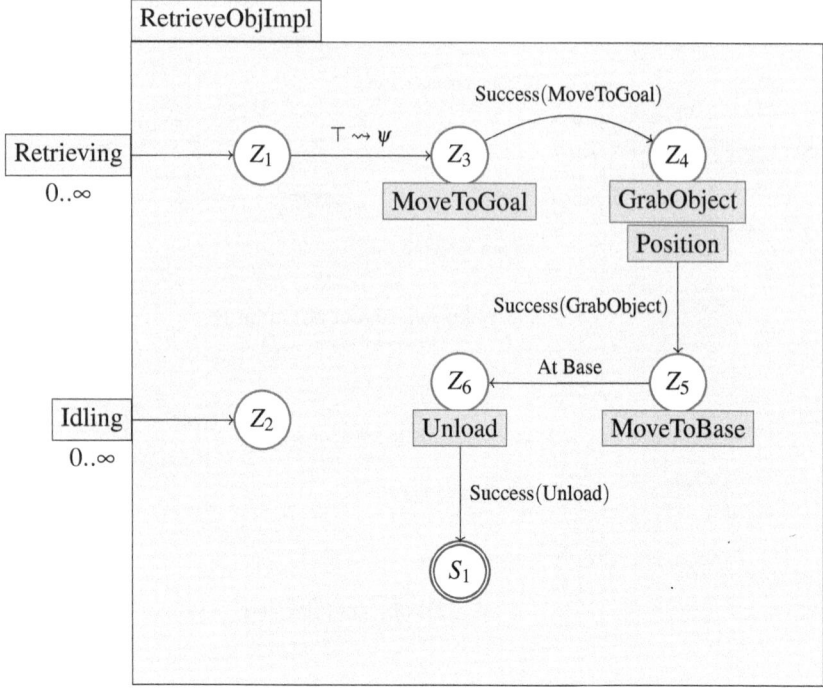

Figure 10.20: Plan for Retrieval

- *Position*(a, p) – Agent a is at position p.

- *Near*(p, o) – Position p is near object o.

- *Component*(o) – o is a component located that is not yet transported.

Figure 10.20 shows a possible realisation of *RetrieveObject*: *RetrieveObjImpl*. It features two tasks, *Retrieving* and *Idling*. The intuition is that surplus trans-porters should idle until more components have been found by the scouts. This is achieved by a combination of runtime condition and utility function:

$$\text{Run}(RetrieveObjImpl) = ComponentsLocated(n)$$
$$\wedge |\{a \mid \text{In}(a, RetrieveObjImpl, Retrieving, z)\}| \leq n$$
$$\mathcal{U}(RetrieveObjImpl) = \text{pri}(RetrieveObjImpl)$$
$$+ \frac{|\{a \mid \text{In}(a, RetrieveObjImpl, Retrieving, z)\}|}{|\{a \mid \text{In}(a, RetrieveObjImpl, \tau, z)\}|}$$

Note that depending on the task allocation algorithm, implicit idling can already achieve the desired behaviour. Here, we use an explicit idling task so that the effect becomes more apparent.

Each transporter that executes the task *Retrieving* should move to a specific object, pick it up and move back to the base with it. This assignment of robots to available objects is done by a constraint system, denoted by ψ.

$$\psi = (\forall a)(\exists g)((\exists z)\, \text{In}(a, RetrieveObjImpl, Retrieving, z)) \rightarrow Position(a, p) \wedge$$
$$GoalPosition(a, g) \wedge Component(o) \wedge Near(g, o) \wedge Distance(p, o, d)$$
$$(\forall a')((\exists z)\, \text{In}(a', RetrieveObjImpl, Retrieving, z) \wedge a \neq a') \rightarrow Position(a', p')$$
$$\wedge Distance(p', o, d') \wedge d < d'$$

where $GoalPosition(a, g)$ is a functional agent fluent. ψ constraints the goal position of an transporter to be close to a component that needs transporting. Furthermore, it constraints the distance to the selected component to be smaller than the distance from any other available transporter to that component. The unfolding step introduced in Definition 8.8 will remove all quantifiers over agents and make the formula applicable for constraint solving.

Thus the behaviour *MoveToGoal* can query for an appropriate position to move to. Once the robots has reached its goal position, it moves onto the state Z_4, in which the two behaviours *GrabObject* and *Position* cooperate to pick up the component. Such a cooperation is achieved by relating the goal position *Position* moves towards with the inverse kinematics problem with which *GrabObject* is confronted. Thereby, *Position* will move towards a position from which the component is graspable.

We omit a detailed description of these behaviours, as their implementation is highly domain-specific and out of scope of this work. The important result here is that the simultaneous collection of objects by multiple robots can easily be described in ALICA. Surplus agents will idle until they can participate. Finally, behaviours controlling different actuators can be combined by relating the corresponding variables using constraints.

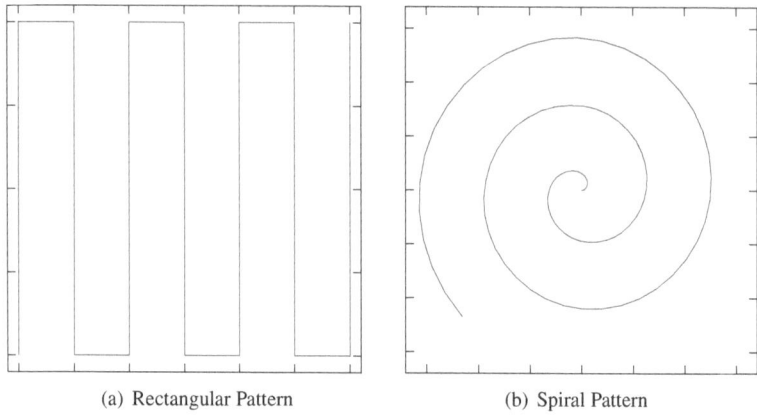

(a) Rectangular Pattern   (b) Spiral Pattern

Figure 10.21: Search Patterns

10.4.2 Exploration

Besides the retrieval of located components, the second important task in this scenario is the search for the components. Cooperative exploration has been extensively researched in the past (e.g., [18]). Efficient cooperative exploration relies on the exchange and merging of data structures representing the environment. This is not a topic covered by ALICA. Any realisation of the plantype *Search* therefore strongly relies on an external domain-specific data representation. Here, we do not go into detail about such representations. Instead, we focus on how constraints can be used to coordinate the movement of multiple robots in order to sweep an area.

Whenever a specific object is searched for, there might be some prior knowledge available indicating where that object is likely to be located. Such prior knowledge might be represented as a probability function or a belief function in the Dempster-Shafer theory [131, 149]. The search pattern used by a team of robots to sweep an area can accommodate for the specifics of such a function. If there is little information available, the probability function approaches a uniform distribution. In this case, a search pattern as depicted in 10.21(a) may be a reasonable choice. If, one the other hand, the probability function is denser in a certain region, e.g., if modelled by a normal distribution, a spiral pattern (10.21(b)) starting at the centre of the normal distribution would be more appropriate.

Following such a search pattern in a formation can be formulated in a simple extension of ALICA constraint systems, namely by connecting subsequent constraint

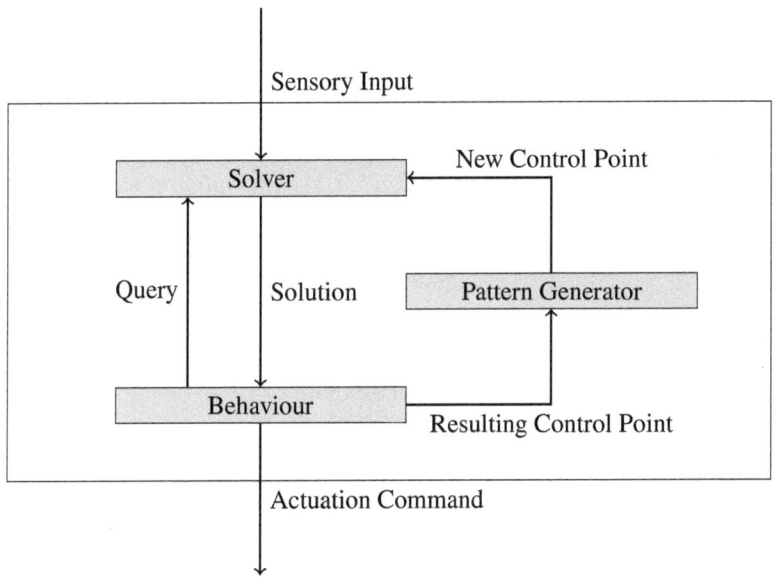

Figure 10.22: Closed Loop Controlling a Formation

problems with a state-full feedback. Here, the formation, such as a line, is described using a point and an angle. The point is used to update a controller, which implements a search pattern. In subsequent problems, the controller provides the corresponding next point, thus enforcing movement along the implemented path.

The proposed architecture is depicted in Figure 10.22. The behaviour queries the solver for a solution to its variables, issues actuator commands and updates the pattern generator with the new control point. The pattern generator fits the control point to its pattern, and presents a new control point, updated by a small interval to the constraint solver for the next iteration. The constraint problem that implements the coordinated movement is defined by the following macro:

$$Line(r,x,y,\alpha,p,\tau,c_x,c_y,\beta) \overset{def}{=} \sqrt{(c_x-x)^2 + (c_y-y)^2} < tol_d \wedge$$

$$\sqrt{(\cos(\beta)-\cos(\alpha))^2 + (\sin(\beta)-\sin(\alpha))^2} < tol_a \wedge$$

$$n = |\{a \mid \mathrm{In}(a,p,\tau,z)\}| \wedge len = 2 \cdot r \cdot n \wedge$$

$$(\forall a)(\exists z)\,\mathrm{In}(a,p,\tau,z) \to (GoalPosition(a,g) \wedge$$

$$|(g_x-c_x)\cos(\beta) + (g_y-c_y)\sin(\beta)| < \varepsilon \wedge$$

$$|g-c| \leq \frac{len}{2} \wedge$$

$$(\forall a')(\exists z)\,\mathrm{In}(a',p,\tau,z)a \neq a' \to$$

$$\left(GoalPosition(a',g') \wedge |g-g'| \geq \frac{len}{n-1} - \varepsilon \right) \bigg)$$

Where c_x, c_y, and β are constrained variables. $Line(n,r,x,y,\alpha,p,\tau,c_x,c_y,\beta)$ enforces the goal positions of all robots executing task τ in plan p to be equally distributed on a line of length $2 \cdot r \cdot n$, where n is the number of robots involved and r their observation radius. The line's position and orientation is specified by the coordinates x and y and the angle α, to which it is orthogonally aligned. The parameters p and τ refer to the agents' plan and task, respectively. Note that this constraint makes use of non-linear and transcendental functions to position the line with respect to the control point (x,y,α).

The constants tol_a, tol_d, and ε implement tolerances. These tolerance provide the necessary leeway to coordinate the CSP without any central control. In order to stabilise positions within the line and to incorporate feedback from the actual positions of the robots, we extend the CSP with an objective function:

$$o(p,\tau) = |\{a \mid \mathrm{In}(a,p,\tau,z)\}| \cdot 10^{12}$$

$$- \sum_{a:\,(\exists z)\,\mathrm{In}(a,p,\tau,z)} |GoalPosition(a) - Position(a)|^2$$

$o(p,\tau)$ reflects the cost to move to the goal position as sum of the squared distances.

The solution to this constraint optimisation problem is two-fold, firstly it contains the target positions of all robots involved, and secondly it contains a derived control point given by c_x, c_y, and β, which deviates from the input control point at most as far as the tolerance levels allow.

Figure 10.23 shows the resulting paths of four simulated robots following a rectangular pattern (10.23(a)) and a spiral search pattern (10.23(b)). The robots

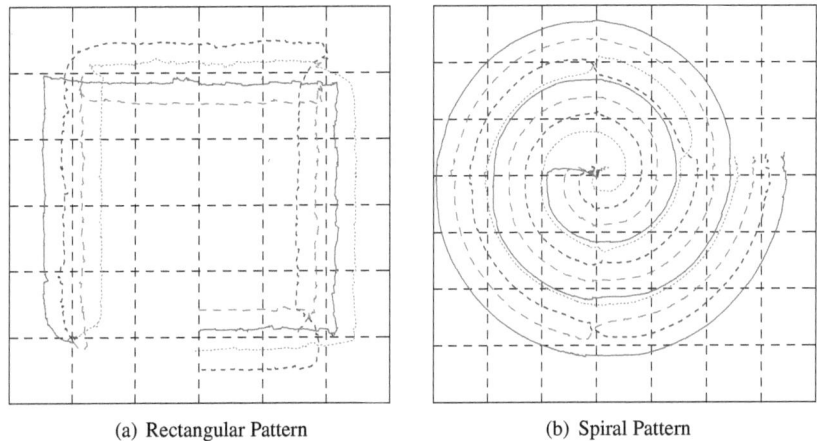

(a) Rectangular Pattern   (b) Spiral Pattern

Figure 10.23: Robots Following Search Patterns

are able to coordinate their movements to consistently form a search line. Note that no additional coordination metaphors are used. The initial state of the pattern generator is determined by the initial solution to the constraint problem. However, the resulting movement is not perfect, the graphs show some noise and in case of the spiral pattern, the robots occasionally switch positions within the formation.

Although these results are not ideal, we could show that formations such as sweep lines along a given pattern can be formulated and coordinated using constraint optimisation problems in ALICA together with a feedback system which generates the search pattern. This approach works only as a proof of concept and needs further study. However, such an analysis is out of scope of this work. The advantages of this approach are clear:

- It directly integrates with other language elements. Thus, one can for instance switch back and forth between different behaviours using transitions or exchange tasks using dynamic reallocation.

- It allows for automatic reformation in case a robot breaks down, in which case the sweep line automatically shrinks, such that no gaps appear in the swept area.

- Finally, the approach has a clear mathematical foundation upon which an analyses can be performed.

10.4.3 Summary

In this section, we used ALICA to model the behaviour of a team of robots which search an unknown area for specific objects, and transport these objects back to a base. We showed how the different modelling elements of ALICA can be combined to form a concise and robust description of the intended team behaviour. This modelling approach is already being used in the DLR-coordinated project IMPERA.

Finally, we sketched how behaviour descriptions based on constraint optimisation can be extended to describe a dynamic system of robots. While the results for the latter are promising, further research in this area is needed, mapping these descriptions onto a theory.

10.5 Rescue Simulation

As a final evaluation scenario, we turn to the rescue domain. The purpose of the following experiments is to

- show that ALICA is usable in other problem domains,

- assess its competitiveness with respect to other techniques,

- evaluate the scalability of ALICA with respect to the number of agents in a team.

In the following, we use RMASBENCH by Kleiner et al. [86], an open-source simulator based on the RoboCup Rescue Simulation Project[1]. In our experiments, a catastrophic event is simulated. Multiple fires are ignited within a city, spread from building to building and threaten to destroy the whole city. The agents in this scenario are fire brigades able to move about and extinguish fires. Figure 10.24 shows a screenshot of the simulation.

One of the main problems the agents are confronted with is to decide which agent is extinguishing which fire. The domain is dynamic, so as the fires spread and are extinguished, the agents need to continuously update their decisions. RMAS-BENCH features some integrated techniques targeting this problem:

- SampleAgents – This strategy acts as a baseline. Each agent greedily selects a nearby target based on the utility function. Since agents make decisions on their own, it can be considered as a simple distributed method.

---

[1]   http://www.robocuprescue.org

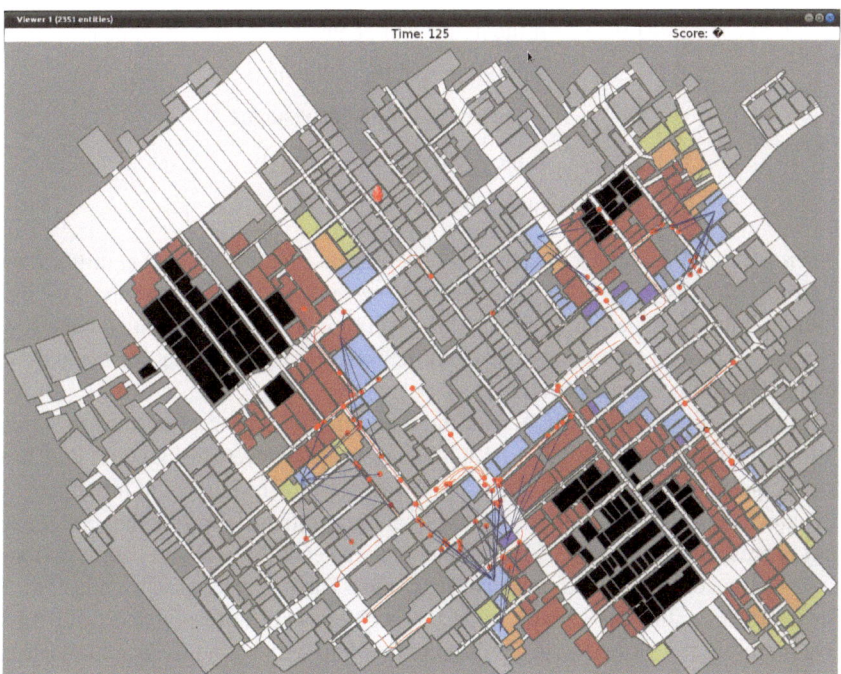

Figure 10.24: RMASBENCH Screenshot – Red dots denote fire brigades, fire of different levels are indicated by yellow, orange, and red buildings. Extinguished buildings are blue and purple. Black buildings are destroyed.

- Hungarian Assignment [90] is a central algorithm for optimising the assignment of n agents to m tasks. It assigns at most one agent per task, and thus will not find the optimal solution in many cases.

- DSA – Distributed Stochastic Algorithm [185] is a decentralized method for distributed constraint optimisation. Each agent calculates the best possible improvement for its own assignment given its current beliefs about the other agents and applies it with a fixed probability (of 0.5 in our experiments). If an agent changes its assignment, it broadcasts the result to the team.

Besides the algorithm to solve this assignment problem, the representation of the problem has a major impact on the performance of the team. The default problem representation used in RMASBENCH associates a utility value with every agent-fire pair:

$$U(a,f) = \begin{cases} 0 & \text{if the number of agents assigned to } f \text{ exceeds} \\ & \text{a building specific value } n(f) \\ \frac{10^{10}}{d(a,f)} & \text{if the fire is still in early stages} \\ \frac{10^3}{d(a,f)} & \text{if the fire burns very strongly} \\ \frac{10^7}{d(a,f)} & \text{otherwise} \end{cases}$$

where $d(a,f)$ denotes the distance between agent a and fire f. This utility puts a strong on fires in early stages, which are easier to extinguish and are typically found on the fringe of larger fires. The resulting matrix, which contains one row per agent and one column per fire is normalised and used as input for the assignment algorithms mentioned above.

In order to compare ALICA with these approaches, we want to transform this utility function into a suitable ALICA concept. However, tasks in ALICA are static, that is, a given plan has a fixed set of tasks, while the number of fires changes dynamically over time. Without an additional generative algorithm, the problem is not expressible as an ALICA task allocation. Instead, the problem can be formulated using constraints and objective functions. The problem class we considered so far uses continuous values, while the problem here is formulated discretely. We therefore transform the given discrete problem into a continuous optimisation problem. The basic idea is to optimise the target positions of the agents. If the distance between a target position and a fire is below a small threshold, the corresponding agent is considered to be assigned to the fire. For this problem, we again use a functional agent fluent for the goal position of an agent, similar to the experiment in Section 10.4.

Let F be the set of fires burning and let $U_s(a,f)$ be the original utility function $U(a,f)$ without the case for too many agents:

$$U_s(a,f) = \begin{cases} \frac{10^{10}}{d(a,f)} & \text{if } f \text{ is still in early stages} \\ \frac{10^3}{d(a,f)} & \text{if } f \text{ burns very strongly} \\ \frac{10^7}{d(a,f)} & \text{otherwise} \end{cases}$$

Then we can formulate a continuous objective function $o(p,\tau)$:

$$o(p,\tau) = \sum_{a:\text{In}(a,p,\tau,z)} \sum_{f\in F} (d_g(a,f) < \varepsilon ? U_s(a,f) : 0)$$

where

- $d_g(a, f)$ denotes the distance between the constrained goal position of agent a and fire f as opposed to the actual distance between a and f, $d(a, f)$.

- The ternary operator ? is used to reify constraints:

$$(\phi ? a : b) \overset{def}{=} \begin{cases} a & \text{if } T(\phi) > 0 \\ b & \text{otherwise} \end{cases}$$

$$\frac{\delta}{\delta x}(\phi ? a : b) \overset{def}{=} \begin{cases} -(a-b)\frac{\delta}{\delta x}T(\neg\phi) & \text{if } T(\phi) > 0 \\ (a-b)\frac{\delta}{\delta x}T(\phi) & \text{otherwise} \end{cases}$$

Reification of constraints is a common constraint programming method to avoid combinatorial explosions in constraints. Here, we defined a special gradient, such that the solver is not confronted with a flat landscape.

The requirement that only a limited number of agents is assigned to each fire can be expressed with the constraint ψ:

$$\psi = \bigwedge_{i=0}^{n} \left(\sum_{a:\ln(a,p,\tau,z)} (d_g(a, f_i) < \varepsilon ? 1 : 0) \right) \leq n(f_i)$$

We do not impose a constraint that requires all agents to be assigned to a fire, since this would lead to an unsatisfiable problem in cases with few fires burning. With the constraint ψ and the objective function $o(p, \tau)$, we have a close approximation of the original assignment problem in a continuous space. Each agent's goal position is a two-dimensional vector in Cartesian space. Suitable bounds for these positions can be derived from the size of the city.

An ALICA behaviour can now query for its goal position, and issue a command to the simulator to move towards and extinguish the fire closest to this position.

The resulting ALICA plan is fairly simple. All agents execute the same task and inhabit the same state. For the purpose of this evaluation, we coin this strategy *GlobalCOP*. The size of the formulae in the COP central to this strategy scales linear with both the number of fires and the number agents. Additionally, the size of the search space grows exponentially with the number of agents. We therefore expect this strategy to fail to scale above a certain amount of agents. However, ALICA has a strong emphasis towards problem decomposition. Exploiting this, we will evaluate a second strategy, called *RegionCOP*. RegionCOP divides the team using four tasks, one for each quadrant of the map. Within each task, the agents solve a COP similar to the one above, but limited to the agents in the corresponding task and the fires in the corresponding quadrant. For task allocation, we use the following utility summand:

$$f(B) = \sum_{i=1}^{i=4} P_i \left(0.9 + \frac{0.1}{|\mathcal{A}|} \sum_{a:\ln(a,p,\tau_i,z)} \left(1 - d(a,c_i)^2 / maxDist^2 \right) \right)$$

where

- $P_i = \min(1, 1 - F_i + \frac{\{a|\ln(a,p,\tau_i,z)\}|}{|\mathcal{A}|})$

- $F_i = \frac{\sum_{f \in M_i} w(f)}{\sum_{f \in F} w(f)}$

- $w(f) = \begin{cases} 9 & \text{if } f \text{ is still in early stages} \\ 1 & \text{if } f \text{ burns very strongly} \\ 4 & \text{otherwise} \end{cases}$

- M_i refers to the set of fires in quadrant i.

- $d(p,q)$ is the usual distance function,

- c_i denotes the centre of quadrant i.

- τ_i is the task associated with quadrant i.

Intuitively, $f(B)$ measures the weighted distribution of fires in the four quadrants and prefers similar distribution of agents. Additionally, agents have a preference towards close quadrants. Due to the large search space, e.g., for four quadrants and 50 agents, 4^{50} possibilities, 5^{50} if idling is allowed, the heuristic has to be very precise. We use a simplification based on a greedy assignment to calculate the heuristic of $f(B)$.

The strategy RegionCOP uses a flat task hierarchy to simplify the constraint optimisation problem. It also forces the agents to work on multiple fire sources at once. In the following experiment we will use it to examine the scalability of task allocation within ALICA.

All experiments use the same scenario: in the city depicted in Figure 10.24, fires are initiated at three different locations. In each experiment, we vary the number of agents available and the time at which the agents are allowed to start. The later the agents are allowed to start, the more difficult it is for them to extinguish all fire sources, since the fires had time to spread. Therefore, we set the start time to $\max(100, 2n + 1)$, where n is the number of agents available. Thus, larger teams are confronted with harder problems. The start time is capped at 100, so that parts of the city still remain before the agents start to act. Additionally, this fixed start value allows us to evaluate how the different strategies utilise additional agents to solve the same problem.

In each time tick of the simulation, each agent makes a single deliberation cycle consisting of the following steps:

- Information from the simulator are integrated into the world model. This information consists of the agent's position and the state of all fires in the city.

- Information about the agent's internal state and its own position is sent to the team.

- An ALICA rule application step takes place, all applicable rules are executed.

- The behaviour queries for the constrained goal position and issues a corresponding command to the simulator.

The time the constraint solver can consume in each tick is set to 600 ms. Recall that this is a soft constraint, the solver will always do at least one run starting from the preferred cluster centre and one exploratory run starting from a random point. The experiments were conducted using nine Intel® Core™i7 CPUs (2.8 GHz) running Linux 3.0.0.

Figure 10.25 shows the experimental results in terms of burned buildings at the end of the simulation, i.e., after all fires are extinguished, or a limit of 300 ticks passed. Each point in the plot is averaged over ten experiments. Note that the performance of GlobalCOP is equivalent to the DSA algorithm up until 40 agents, at which point GlobalCOP starts to perform worse than DSA. The strategy modification RegionCOP performs best in this scenario, although there is little difference to DSA at higher agents numbers. The other assignment strategies perform much worse.

Furthermore, we see that the number of burned buildings increase for all strategies up until 50 agents due to the delayed start of the agents. However, for the experiments with fixed start time, with 50 agents and above, additional agents only improve the performance slightly, regardless of the algorithm.

We conclude that the continuous version of the original assignment problem together with the ALICA solver leads to a comparable performance. Note however, that the DSA approach is much less computational expensive than the continuous solver in ALICA. ALICA agents always consider the whole constraint problem, while DSA controlled agents only consider the choices the local agent has. Moreover, the search space presented to the ALICA solver is much larger, due to the

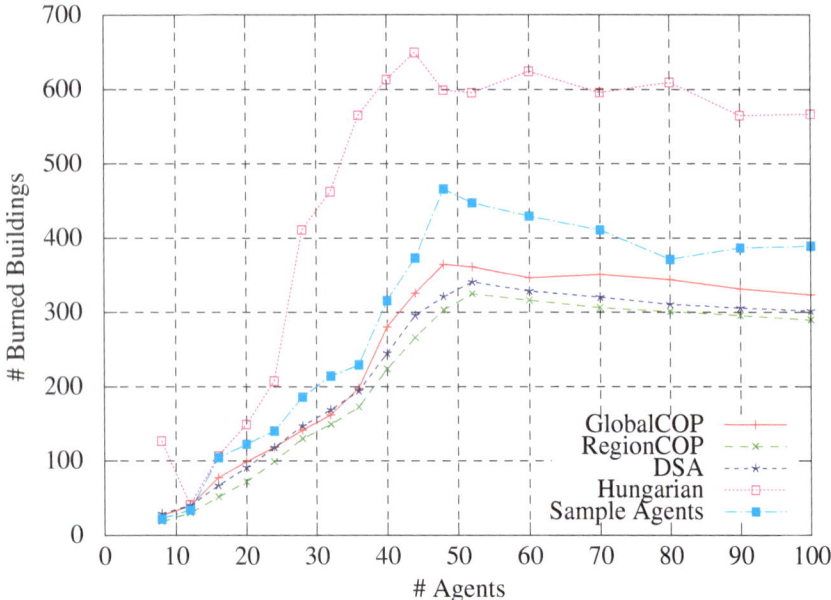

Figure 10.25: Performance in the Fire Extinguish Scenario: Burned Buildings vs. Number of Agents

mapping to a continuous problem. Our approach is geared towards such continuous domains, and should be more flexible with respect to the problem if continuous values are essential.

The scalability of the COP solver can be assessed based on Figure 10.26. It shows the average number of function evaluations done by each agent in each tick for world states with 60 to 80 fires on a logarithmic scale. The number of evaluations decreases drastically with the number of agents. This is due to the size of the formulae increasing, the dimensionality of the problem increasing, and the number of simulated agents increasing, which have to share the available computing power. At higher agent numbers, the solvers often violate the soft constraint of 600 ms solving time, thus the number of function evaluations levels out. The figure also shows that the problem decomposition achieved by RegionCOP allows for an order of magnitude more function evaluations per tick, enabling the agents to search the smaller search spaces more thoroughly. Note that the number of func-

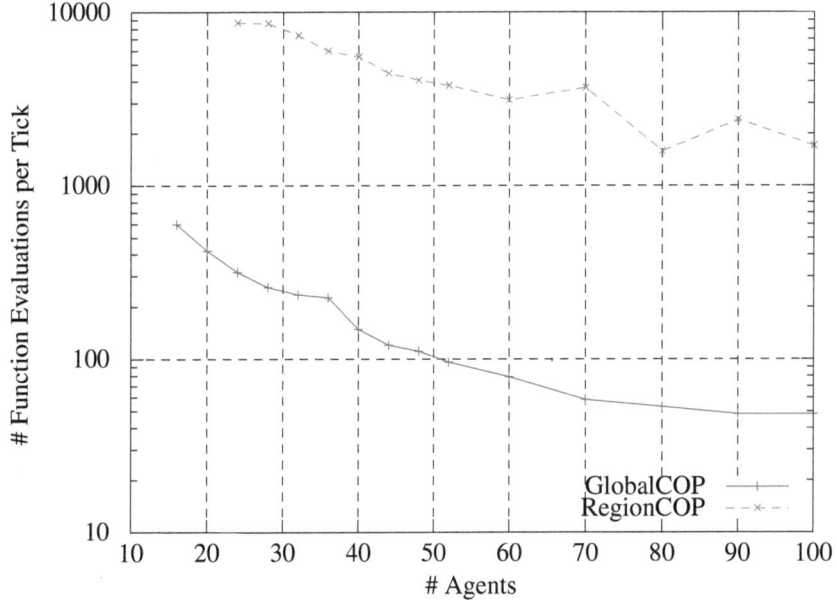

Figure 10.26: Function Evaluations for 60 to 80 Fires

tion evaluations in Figure 10.26 only reflects the number of calculations within the allotted time, not the number of calculations till an assignment was found. The solver spends all available time trying to find a better solution, oblivious to whether or not one exists.

Of course, the improvement in solving speed by RegionCOP does not come for free. It requires the agents to be allocated to four tasks. Figure 10.27 illustrates how this algorithm scales. It shows the number of performed expansion steps by the A* search algorithm. While the average case grows almost linear with the number of agents, the worst case number of steps is exponential in the number of agents involved. This is not surprising, the problem is NP-hard afterall. However, with an average of approximately 200 expansion steps in order to allocate 100 agents, the algorithm scales rather well. This performance highly depends on the specified heuristic.

Note that the performance of the tested approaches differs with the scenario. As an example, Figure 10.28 shows the performance of the different algorithms

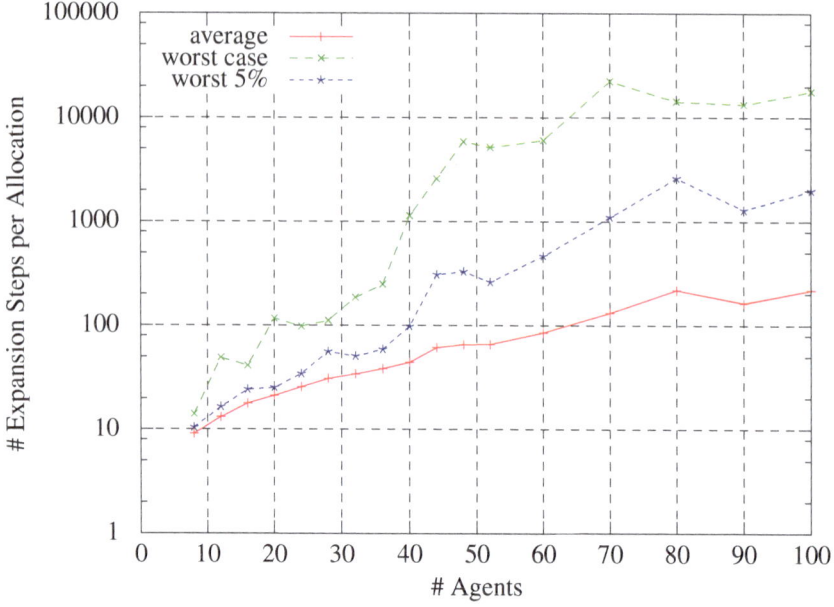

Figure 10.27: Task Allocation Expansion Steps

from the previous experiment in a different scenario. The data is averaged over ten experiments. While RegionCOP exhibits consistently good results, GlobalCOP struggles to keep the fire under control. It was able to contain the fire in four out of ten experiments. All other strategies consistently failed to extinguish all fires by the end of the experiment, i.e., after 300 ticks. Interestingly, the naive SampleAgents strategy achieved better results than DSA in this specific experiment, illustrating the impact of the scenario.

In summary, we presented an evaluation in the rescue domain, and were able to coordinate up to 100 agents using ALICA. We transferred a discrete optimisation problem into a continuous domain and showed that the resulting problem can be solved using the approach presented in this work. Moreover, the corresponding strategy is strikingly similar in performance to the original problem formulation together with the Distributed Stochastic Algorithm up until 40 agents, at which point ALICA performed slightly worse. Finally, we showed that a fairly simple problem decomposition can be implemented using ALICA language elements such as tasks

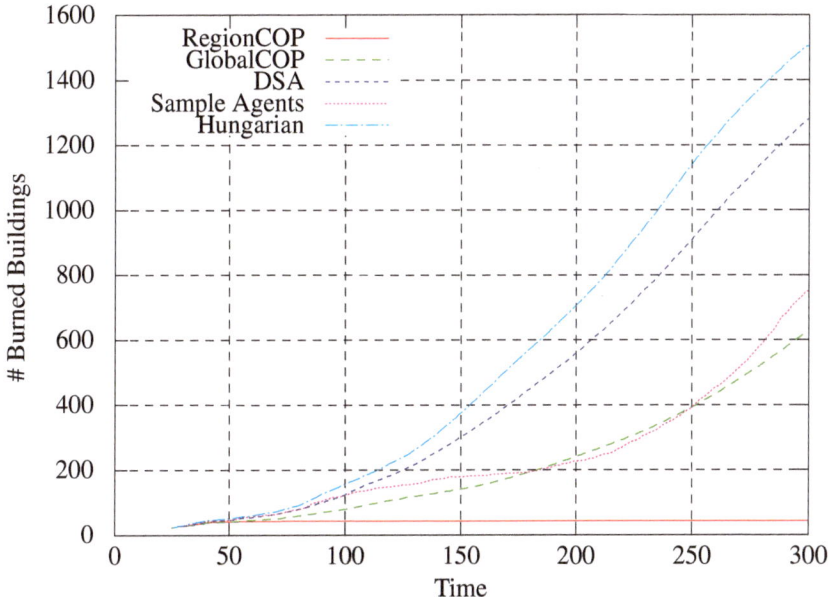

Figure 10.28: Exemplary Fire Fighting Performance with 18 Agents

and utility functions, and we showed that the resulting strategy outperformed the original one.

11 Conclusion

In this thesis, we presented a comprehensive solution for modelling the behaviour of a team of autonomous robots. This solution is geared towards dynamic domains, in which the robots have to exhibit a high degree of reactivity, while acting in a coherent fashion.

The main contribution of this thesis constitutes a language, in which team behaviour can be described from a global perspective. The semantics of this language define an execution layer, which is described in detail. The language abstracts away from concrete robots or agents by using capabilities to map them onto roles. Roles are then used to allocate robots dynamically to tasks within plans. These plans are at the core of the language. Using hierarchies of finite state machines, they describe strategies or recipes to tackle problems. The execution layer features an efficient recursive task allocation algorithm, which is used by each agent to locally compute relevant task allocations. Conflicts are reliably detected and resolved by adaptively switching the decision protocol temporarily and locally to a central assignment algorithm. Incapacitated robots are seamlessly compensated for by the team.

This core language is extended in the third part of this dissertation with the ability to express complex intentions using non-linear constraint satisfaction and optimisation problems over continuous domains. These problems are solved during runtime by the participating robots using cooperating anytime solvers. The cooperation between these solvers emphasizes reactivity of the individual robot, but provides coherence under sensory noise.

The resulting framework is completely distributed and constitutes a novel combination of finite state machines, utility-based decision making, and constraint programming. This combination yields a very powerful modelling language accessible to system designers as well as generative algorithms, such as planners.

On the theoretical level, we obtained results for hierarchical task allocation, which yielded provable conditions that plans have to satisfy in order to allow for conflict free allocations. We provide two different task allocation algorithms with different requirements, one allowing for agents to passively participate in a plan and one which requires all agents to actively commit to a task at all times.

A reference implementation is open source and successfully used by the RoboCup team Carpe Noctem. Furthermore, it is also used in the context of the DLR-coordinated project IMPERA.

The evaluation in the previous chapter showed that ALICA is able to successfully control and coordinate robots in various domains, such as robotic soccer, rescue, and exploration. Furthermore, the experiments demonstrate robustness against sensory noise and unreliable communication. Finally, results drawn from the rescue domain show that the integrated algorithms are competitive with state-of-the-art approaches to task assignment and that ALICA is able to control and coordinate larger teams.

This scalability is achieved by the hierarchical structure of ALICA programs together with the locality principle, the cooperation between the anytime solvers of the agents, and the heuristic functions used to guide the task allocation algorithm. However, for very large sets of agents, additional paradigms might be needed to support concise modelling, such as social laws or norms. The provided constraint-based modelling approach is a promising foundation for an integration of such organisational paradigms.

11.1 Requirements Revisited

In Section 1.2, we introduced a set of challenging domain requirements the presented approach should be able to cope with. Here, we summarise how these are solved.

Continuously Changing Environment Each agent in ALICA acts upon local decisions, and does not require any prior communication, thus it can react directly to any unforeseen changes. Dynamic reallocation allows the agents to switch tasks or even complete plans in order to adapt to such changes. Moreover, the constraint solver always considers the current situation by solving an on-demand constructed problem, thereby emphasising reactivity. On the implementation level this is supported by the rule application component, which performs updates anywhere in the program hierarchy without additional latencies due to synchronisation between threads.

Noisy Sensors While ALICA does not come with a perception or sensor fusion component and should be used in conjunction with corresponding components, it provides features that counteract noise while retaining a certain level of reactivity. Firstly, thresholds and similarity values can be used to stabilise task allocations in the presence of noise. Secondly, the team is able to stabilise solutions to constraint problems without any loss of reactivity for the individual agent, as shown in Section 10.3.

Partially Observable Environment As an execution and coordination layer, ALICA does not directly tackle this issue. A solution to this problem is part of the representation of the environment, which ALICA does not incorporate, in order to maintain domain independence. However, since ALICA agents are exchanging information about what they are doing, the overall behaviour is less dependent on an individual's beliefs. This is most apparently reflected in the locality principle ALICA follows, which entails that an agent does not concern itself with problems or plans it is not actively or passively participating in. Local information is therefore sufficient for an agent to fulfil its part within the team.

Unreliable Communication Robustness under bad network quality was evaluated in detail in Section 10.2. The results show a high degree of coherence even under high packet loss or delay.

Failing Team Members Incapacitated robots are compensated for on the fly by dynamic reallocation. Should the currently executed plan no longer be executable due to missing capabilities or robots, an alternative is selected, if one exists. Otherwise, a failure is raised and propagated through the program hierarchy until it can be resolved.

11.2 Outlook and Future Work

The work presented in this thesis constitutes a comprehensive solution to modelling and coordinating the behaviour of a team of autonomous mobile robots. Still there are some open questions left. Furthermore, ALICA is meant to be extensible such that future components can be integrated easily. In the following we summarise the most important questions and possible extensions.

Integration of Planning ALICA does not feature a planning component in the classical sense. However, the language is designed with planning in mind. Therefore, post conditions can be defined for the appropriate language elements such as terminal states and behaviours. These are not used during runtime, but play a pivotal role during planning. Integration of planning is investigated within the research project IMPERA.

SMT-Based Constraint Solving and Optimisation In Chapter 8, we advocated the use of SMT solvers within ALICA and sketched possible approaches to extend such solvers with the ability to track and coordinate solutions. To the best of our knowledge, currently no SMT solver exists that is competitive with the solver discussed in this work. However, the experiment in Section 9.3 suggests that SMT-based techniques can lead to a tremendous gain in performance.

Merging Task Allocation and Constraint Solving As mentioned in Section 9.7, ALICA agents are confronted with two potentially computational expensive problems during runtime. The first is task (re-)allocation, the second constraint solving. Integrating the two could lead to even more concise descriptions of behaviours and better performance of the team, by allowing tasks to be allocated with respect to constraint problems. This is currently only possible in a limited fashion, namely by referring to the current solutions from within conditions or utility functions. How to deal with the resulting complexity is an open question.

Modelling Dynamic Behaviour using Differential Constraint Problems The provided methodology to formulate behaviour using constraint problems can easily be extended to incorporate a feedback, essentially yielding constraint problems that feature differential equations over time. In Section 10.4, we discussed how this can be used to express dynamic formations without the use of a central component or the specification of an initial state. A theory is still missing that entails how the different properties of such a system such as the velocity of the robots and the degree of coherence can be controlled. Investigating these systems and relating them to other approaches such as the Dual Dynamics design scheme by Jaeger and Christaller [77] is an exciting topic for future work.

Conflict Resolution for CSPs Our approach to provide coherent solutions to the constraint problems currently does not extend to the case where the individual constraint problems each robot considers are significantly different but related through some variables. Tackling this problem requires additional messages to be passed by the solvers involved along the lines of DCOP solvers as discussed by Petcu [121]. The additional challenges introduced by the combination of non-linear constraints, continuous domains, the dynamic environment, and unreliable communication still need to be addressed.

Bibliography

[1] Till Amma, Philipp Baer, Kai Baumgart, Philipp Burghardt, Kurt Geihs, Janosch Henze, Stephan Opfer, Stefan Niemczyk, Roland Reichle, Daniel Saur, Andreas Scharf, Jens Schreiber, Martin Segatz, Florian Seute, Hendrik Skubch, Stefan Triller, Michael Wagner, and Andreas Witsch. Carpe Noctem 2009. In *RoboCup 2009 International Symposium*, Graz, June 2009. TU Graz.

[2] Krzysztof R. Apt and Mark Wallace. *Constraint Logic Programming using Eclipse*. Cambridge University Press, New York, NY, USA, 2007. ISBN 0521866286.

[3] Matthew Arnold, Stephen J. Fink, David Grove, Michael Hind, and Peter F. Sweeney. A survey of adaptive optimization in virtual machines. In *Proceedings of the IEEE Special Issue on Program Generation, Optimization, and Adaptation*, volume 93, pages 449–466, 2005. doi:10.1109/JPROC.2004.840305.

[4] John Aycock. A brief history of just-in-time. *ACM Comput. Surv.*, 35(2): 97–113, 2003. doi:10.1145/857076.857077.

[5] Philipp A. Baer. *Platform-Independent Development of Robot Communication Software*. Phd thesis, University of Kassel, Kassel, 2008. URL http://www.upress.uni-kassel.de/publi/abstract.php?978-3-89958-644-2.

[6] Andreas Bauer, Markus Pister, and Michael Tautschnig. Tool-support for the analysis of hybrid systems and models. In *Design, Automation and Test in Europe (DATE)*, pages 924–929, 2007.

[7] W. Beaton and J. d. Rivieres. Eclipse Platform Technical Overview. Technical report, The Eclipse Foundation, 2006.

[8] R. E. Bellman. *Dynamic Programming*. Princeton University Press, Princeton, N.J., 1957.

[9] Roger Bemelmans, Gert Jan Gelderblom, Pieter Jonker, and Luc De Witte. Socially assistive robots in elderly care: A systematic review into effects and effectiveness. *Gerontechnology Journal*, 8(2):94–103, 2010. doi:10.1016/j.jamda.2010.10.002.

[10] Frédéric Benhamou and Laurent Granvilliers. Continuous and Interval Constraints. In Francesca Rossi, Peter van Beek, and Toby Walsh, editors, *Handbook of Constraint Programming*, pages 571–603. Elsevier, 2006. ISBN 978-0-444-52726-4.

[11] J. Bohren and S. Cousins. The smach high-level executive. *Robotics Automation Magazine, IEEE*, 17(4):18–20, dec 2010. ISSN 1070-9932. doi:10.1109/MRA.2010.938836.

[12] Michael Bratman. *Intentions, Plans, and Practical Reason*. Harvard University Press, 1987.

[13] Michael Brenner. A multiagent planning language. In *Workshop on ICAPS*, May 2003.

[14] Gerhard Brewka, Ilkka Niemelä, and Miroslaw Truszczynski. Preferences and nonmonotonic reasoning. *AI Magazine*, 4:69–78, 2008.

[15] Susan S. Brilliant and Timothy R. Wiseman. The first programming paradigm and language dilemma. In *Proceedings of the twenty-seventh SIGCSE technical symposium on Computer science education*, SIGCSE '96, pages 338–342, New York, NY, USA, 1996. ACM. ISBN 0-89791-757-X. doi:10.1145/236452.236572.

[16] Kenneth N. Brown and Ian Miguel. Uncertainty and Change. In Francesca Rossi, Peter van Beek, and Toby Walsh, editors, *Handbook of Constraint Programming*, pages 731–760. Elsevier, 2006. ISBN 978-0-444-52726-4.

[17] Wray Buntine. Generalized subsumption and its applications to induction and redundancy. *Artificial Intelligence*, 36:149–176, September 1988. ISSN 0004-3702. doi:10.1016/0004-3702(88)90001-X.

[18] Wolfram Burgard, Mark Moors, Cyrill Stachniss, and Frank E Schneider. Coordinated multi-robot exploration. *IEEE Transactions on Robotics*, 21 (3):376–386, 2005. doi:10.1.1.59.4390.

[19] Carlos Caleiro and Ricardo Gonçalves. On the algebraization of many-sorted logics. In *Recent Trends in Algebraic Development Techniques - Selected Papers*, pages 21–36. Springer-Verlag, 2007.

[20] Adam Campbell and Annie S. Wu. Task and role allocation within multi-agent and robotics research. Technical Report 05, UCF, 2007.

[21] Adam Campbell and Annie S. Wu. Multi-agent role allocation: issues, approaches, and multiple perspectives. *Autonomous Agents and Multi-Agent Systems*, 22(2):317–355, 2011. doi:10.1007/s10458-010-9127-4.

[22] D. Challet and Y.-C. Zhang. Emergence of cooperation and organization in an evolutionary game. *Physica A: Statistical Mechanics and its Applications*, 246(3–4):407–418, 1997. ISSN 0378-4371. doi:10.1016/S0378-4371(97)00419-6.

[23] Stuart Chalmers and Peter M.D. Gray. BDI agents and constraint logic. *AISB Journal Special Issue on Agent Technology*, 1(1):21–40, 2001.

[24] Shyamal Suhana Chandra and Kailash Chandra. A comparison of Java and C#. *Journal of Computing Sciences in Colleges*, 20:238–254, February 2005. ISSN 1937-4771.

[25] Antonio Chella, Massimo Cossentino, Roberto Pirrone, and Andrea Ruisi. Modeling ontologies for robotic environments. In *Proceeding of the 14th International Conference on Software Engineering and Knowledge Engineering*, pages 15–19, 2002.

[26] Xinguang Chen and Peter van Beek. Conflict-directed backjumping revisited. *Journal of Artificial Intelligence Research (JAIR)*, 14:53–81, 2001. doi:10.1613/jair.788.

[27] David Cohen and Peter Jeavons. The Complexity of Constraint Languages. In Francesca Rossi, Peter van Beek, and Toby Walsh, editors, *Handbook of Constraint Programming*, pages 245–280. Elsevier, 2006. ISBN 978-0-444-52726-4.

[28] Philip R. Cohen and Hector J. Levesque. Intention is choice with commitment. *Artificial Intelligence*, 42(2-3):213–261, 1990. ISSN 0004-3702. doi:10.1016/0004-3702(90)90055-5.

[29] Stephen A. Cook. The complexity of theorem-proving procedures. In *Proceedings of the third annual ACM symposium on Theory of computing*, STOC '71, pages 151–158, New York, NY, USA, 1971. ACM. doi:10.1145/800157.805047.

[30] James M. Crawford and Larry D. Auton. Experimental results on the crossover point in random 3-sat. *Artificial Intelligence*, 81(1-2):31–57, 1996.

[31] Nadia Creignou, Sanjeev Khanna, and Madhu Sudan. *Complexity classifications of boolean constraint satisfaction problems.* Society for Industrial and Applied Mathematics, Philadelphia, PA, USA, 2001. ISBN 0-89871-479-6.

[32] Robert J. Dakin. A tree-search algorithm for mixed integer programming problems. *The Computer Journal*, 8(3):250–255, March 1965. ISSN 1460-2067. doi:10.1093/comjnl/8.3.250.

[33] Aniruddha Dasgupta and Aditya K. Ghose. CASO: A Framework for dealing with objectives in a constraint-based extension to AgentSpeak(L). In Vladimir Estivill-Castro and Gillian Dobbie, editors, *Twenty-Ninth Australasian Computer Science Conference (ACSC 2006)*, volume 48 of *CR-PIT*, pages 121–126, Hobart, Australia, 2006. ACS.

[34] Mehdi Dastani, M. Birna, Riemsdijk Frank Dignum, and John-Jules Ch. Meyer. A programming language for cognitive agents: Goal directed 3APL. In *Programming Multi-Agent Systems, First International Workshop, PRO-MAS 2003, Melbourne, Australia*, pages 111–130. Springer, July 2003.

[35] Mehdi Dastani, Dirk Hobo, and John-Jules Ch. Meyer. Practical extensions in agent programming languages. In *AAMAS '07: Proceedings of the 6th international joint conference on Autonomous agents and multiagent systems*, pages 1–3, New York, NY, USA, 2007. ACM. ISBN 978-81-904262-7-5.

[36] Martin Davis and Hilary Putnam. A computing procedure for quantification theory. *Journal of the ACM*, 7:201–215, July 1960. ISSN 0004-5411. doi:10.1145/321033.321034.

[37] Giuseppe de Giacomo, Yves Lespérance, and Hector J. Levesque. Congolog, a concurrent programming language based on the situation calculus. *Artificial Intelligence*, 121(1-2):109–169, August 2000. ISSN 0004-3702. doi:10.1016/S0004-3702(00)00031-X.

[38] Joris De Schutter, Tinne De Laet, Johan Rutgeerts, Wilm Decré, Ruben Smits, Erwin Aertbeliën, Kasper Claes, and Herman Bruyninckx. Constraint-based task specification and estimation for sensor-based robot systems in the presence of geometric uncertainty. *International Journal of Robotics Research*, 26:433–455, May 2007. ISSN 0278-3649. doi:10.1177/027836490707809107.

[39] Wilm Decré, Ruben Smits, Herman Bruyninckx, and Joris De Schutter. Extending iTaSC to support inequality constraints and non-instantaneous task specification. In *Proceedings of the 2009 IEEE International Conference on Robotics and Automation*, ICRA'09, pages 964–971, Piscataway, NJ, USA, 2009. IEEE Press. ISBN 978-1-4244-2788-8.

[40] Gilles Dowek, César Muñoz, and Corina Păsăreanu. A small-step semantics of PLEXIL. Technical Report 2008-11, National Institute of Aerospace, Hampton, VA, 2008.

[41] Markus Eich and Frank Kirchner. Reasoning about geometry: An approach using spatial-descriptive ontologies. In *Workshop AILog 19th European Conference on Artificial Intelligence ECAI10, Lisbon*, 2010.

[42] Ronald Fagin, Joseph Y. Halpern, Yoram Moses, and Moshe Y. Vardi. *Reasoning About Knowledge*. MIT Press, 1995. ISBN 0262061627.

[43] Tomás Feder and Moshe Y. Vardi. The computational structure of monotone monadic snp and constraint satisfaction: A study through datalog and group theory. *SIAM J. Comput.*, 28(1):57–104, February 1999. ISSN 0097-5397. doi:10.1137/S0097539794266766.

[44] K. Feher. *Telecommunications Measurements, Analysis, and Instrumentation*. Noble Publishing classic series. Noble Publishing, 1996. ISBN 9781884932038.

[45] Richard Fikes and Nils J. Nilsson. Strips: A new approach to the application of theorem proving to problem solving. In *Proceedings of the 2nd International Joint Conference on Artificial intelligence (IJCAI)*, pages 608–620, San Francisco, CA, USA, 1971. Morgan Kaufmann Publishers Inc.

[46] FIPA. *FIPA ACL Message Structure Specification*. FIPA, 2001. URL http://www.fipa.org/specs/fipa00061/. (accessed 2012-05-23).

[47] FIPA. *FIPA Communicative Act Library Specification*. FIPA, December 2002. URL http://www.fipa.org/specs/fipa00037/. (accessed 2012-05-23).

[48] Martin Fränzle, Christian Herde, Tino Teige, Stefan Ratschan, and Tobias Schubert. Efficient solving of large non-linear arithmetic constraint systems with complex boolean structure. *Journal on Satisfiability, Boolean Modeling and Computation*, 1:209–236, 2007.

[49] Thom Frühwirth. Introducing simplification rules. Technical Report ECRC-LP-63, European Computer-Industry Research Centre, München, Germany, October 1991. Presented at the Workshop Logisches Programmieren, Goosen/Berlin, Germany, and the Workshop on Rewriting and Constraints, Dagstuhl, Germany,.

[50] Thom Frühwirth. Theory and practice of constraint handling rules. *The Journal of Logic Programming*, 37(1–3):95–138, 1998. ISSN 0743-1066. doi:10.1016/S0743-1066(98)10005-5. URL http://www.sciencedirect.com/science/article/pii/S0743106698100055.

[51] D. Gale. *The Theory of Linear Economic Models*. Economics / mathematics. University of Chicago Press, 1989. ISBN 9780226278841. URL http://books.google.de/books?id=3t3F9rLAZnYC.

[52] J. H. Gallier. *Logic for Computer Science: Foundations of Automatic Theorem Proving*, chapter 10. Many-Sorted First-Order Logic. Harper & Row Publishers, Inc., 1985. Out of print, available via www.cis.upenn.edu/~cis610/logic.pdf.gz (last accessed 10-12-2011).

[53] Harald Ganzinger, George Hagen, Robert Nieuwenhuis, Albert Oliveras, and Cesare Tinelli. Dpll(t): Fast decision procedures. In Rajeev Alur and Doron Peled, editors, *Computer Aided Verification, 16th International Conference, CAV 2004, Boston, MA, USA, July 13-17, 2004, Proceedings*, volume 3114 of *Lecture Notes in Computer Science*, pages 175–188. Springer, 2004. ISBN 3-540-22342-8. doi:10.1007/978-3-540-27813-9_14.

[54] H. Garcia-Molina. Elections in a distributed computing system. *IEEE Transactions on Computers*, 31:48–59, January 1982. ISSN 0018-9340. doi:10.1109/TC.1982.1675885.

[55] Brian P Gerkey. A formal analysis and taxonomy of task allocation in multi-robot systems. *The International Journal of Robotics Research*, 23(9):939–954, 2004. doi:10.1177/0278364904045564.

[56] Jens Gerlach and Joachim Kneis. Generic Programming for Scientific Computing in C++, Java™ and C#. In Xingming Zhou, Ming Xu, Stefan Jähnichen, and Jiannong Cao, editors, *Advanced Parallel Processing*

Technologies, volume 2834 of *Lecture Notes in Computer Science*, pages 301–310. Springer Berlin / Heidelberg, 2003. ISBN 978-3-540-20054-3. doi:10.1007/978-3-540-39425-9_37.

[57] Giuseppe Giacomo, Yves Lespérance, Hector J. Levesque, and Sebastian Sardina. IndiGolog: A High-Level Programming Language for Embedded Reasoning Agents. In Amal El Fallah Seghrouchni, Jürgen Dix, Mehdi Dastani, and Rafael H. Bordini, editors, *Multi-Agent Programming Languages, Tools and Applications*, volume 1, chapter 2, pages 31–72. Springer, 2009. doi:10.1007/978-0-387-89299-3_2.

[58] Piotr J. Gmytrasiewicz and Edmund H. Durfee. Decision-theoretic recursive modeling and the coordinated attack problem. In *Proceedings of the first international conference on Artificial intelligence planning systems*, pages 88–95, San Francisco, CA, USA, 1992. Morgan Kaufmann Publishers Inc. ISBN 1-55860-250-X. URL http://dl.acm.org/citation.cfm?id=139492.139503.

[59] Alexandre Goldsztejn, Claude Michel, and Michel Rueher. An efficient algorithm for a sharp approximation of universally quantified inequalities. In Roger L. Wainwright and Hisham Haddad, editors, *Proceedings of the 2008 ACM Symposium on Applied Computing (SAC), Fortaleza, Ceara, Brazil, March 16-20, 2008*, pages 134–139. ACM, 2008. ISBN 978-1-59593-753-7. doi:10.1145/1363686.1363724.

[60] Alexandre Goldsztejn, Claude Michel, and Michel Rueher. Efficient handling of universally quantified inequalities. *Constraints*, 14:117–135, March 2009. ISSN 1383-7133. doi:10.1007/s10601-008-9053-0.

[61] Robin Gras, Didier Devaurs, Adrianna Wozniak, and Adam Aspinall. An individual-based evolving predator-prey ecosystem simulation using a fuzzy cognitive map as the behavior model. *Artificial Life*, 15(4):423–463, 2009. doi:10.1162/artl.2009.Gras.012.

[62] Jim Gray. Notes on data base operating systems. In *Operating Systems, An Advanced Course*, pages 393–481, London, UK, 1978. Springer-Verlag. ISBN 3-540-08755-9. URL http://dl.acm.org/citation.cfm?id=647433.723863.

[63] Barbara J. Grosz and Sarit Kraus. Collaborative plans for complex group action. *Artificial Intelligence*, 86:269–357, 1996.

[64] Barbara J. Grosz and Candace L. Sidner. Plans for discourse. In P. R. Cohen, J. Morgan, and M. E. Pollack, editors, *Intentions in Communication*, chapter 20, pages 417–444. MIT Press, Cambridge, MA, 1990.

[65] Object Management Group. Data distribution service (dds) specification v1.2. http://www.omg.org/spec/DDS/1.2/ (accessed 2012-05-22), 2007.

[66] Nikolaus Hansen and Andreas Ostermeier. Adapting arbitrary normal mutation distributions in evolution strategies: The covariance matrix adaptation. In *International Conference on Evolutionary Computation*, pages 312–317, 1996.

[67] P. E. Hart, N. J. Nilsson, and B. Raphael. A formal basis for the heuristic determination of minimum cost paths. *Systems Science and Cybernetics, IEEE Transactions on*, 4(2):100–107, 1968. doi:10.1109/TSSC.1968.300136.

[68] W. K. Hartmann. PLANETARY SCIENCE: The Shape of Kleopatra. *Science*, 288:820–821, 2000. doi:10.1126/science.288.5467.820.

[69] Koen Hindriks, Frank S. De Boer, Wiebe Van Der Hoek, and John jules Ch. Meyer. A Formal Embedding of AgentSpeak(L) in 3APL. Technical report, Advanced Topics in Artificial Intelligence, Springer Verlag LNAI 1502, 1998.

[70] Koen V. Hindriks and John-Jules Ch. Meyer. Agent Logics as Program Logics: Grounding KARO. In Christian Freksa, Michael Kohlhase, and Kerstin Schill, editors, *KI*, volume 4314 of *Lecture Notes in Computer Science*, pages 404–418. Springer, 2006.

[71] Koen V. Hindriks, Frank S. De Boer, Wiebe Van Der Hoek, and John-Jules Ch. Meyer. Agent programming in 3APL. *Autonomous Agents and Multi-Agent Systems*, 2(4):357–401, 1999. ISSN 1387-2532.

[72] Steffen Hölldobler and Josef Schneeberger. A new deductive approach to planning. *New Generation Computing*, 8(3):225–244, 1990. doi:10.1007/BF03037518.

[73] Bryan Horling and Victor Lesser. A Survey of Multi-Agent Organizational Paradigms. *The Knowledge Engineering Review*, 19(4):281–316, 2005. URL http://mas.cs.umass.edu/paper/366.

[74] Bryan Horling, Victor Lesser, Regis Vincent, Tom Wagner, Anita Raja, Shelley Zhang, Keith Decker, and Alan Garvey. The TAEMS White Paper, 1999. URL http://mas.cs.umass.edu/paper/182.

[75] Marcus Huber and Edmund H. Durfee. On acting together: Without communication. In *Working Notes of the AAAI Spring Symposium on Representing Mental States and Mechanisms*, pages 60–71, 1995.

[76] *Information technology - XML Metadata Interchange (XMI) – ISO/IEC 19503:2005-11*. International Organization for Standardization, 2005.

[77] Herbert Jaeger and Thomas Christaller. Dual Dynamics: Designing Behavior Systems for Autonomous Robots. *Artificial Life and Robotics*, 2:76–79, 1998.

[78] Nicholas R. Jennings. Controlling cooperative problem solving in industrial multi-agent systems using joint intentions. *Artificial Intelligence*, 75(2): 195–240, 1995.

[79] Richard E. Jones and Rafael Dueire Lins. *Garbage Collection: Algorithms for Automatic Dynamic Memory Management*. John Wiley, 1996. ISBN 0-471-94148-4.

[80] Ari K. Jónsson and Jeremy Frank. A framework for dynamic constraint reasoning using procedural constraints. In Werner Horn, editor, *ECAI 2000, Proceedings of the 14th European Conference on Artificial Intelligence, Berlin, Germany, August 20-25, 2000*, pages 93–97. IOS Press, 2000.

[81] Gal A. Kaminka and Inna Frenkel. Flexible teamwork in behavior-based robots. In Manuela M. Veloso and Subbarao Kambhampati, editors, *Proceedings of The Twentieth National Conference on Artificial Intelligence and the Seventeenth Innovative Applications of Artificial Intelligence Conference, July 9-13, 2005, Pittsburgh, Pennsylvania, USA*, pages 108–113. AAAI Press / The MIT Press, 2005. ISBN 1-57735-236-X.

[82] Gal A. Kaminka and Milind Tambe. Robust agent teams via socially-attentive monitoring. *Journal of Artificial Intelligence Reasearch*, 12:105–147, 2000. doi:10.1613/jair.682.

[83] Hirofumi Katsuno and Alberto O. Mendelzon. On the difference between updating a knowledge base and revising it. In *Proceedings of the Second International Conference on the Principles of Knowledge Representation and Reasoning (KR91)*, pages 387–394. Morgan Kaufmann, 1991.

[84] Kristian Kersting, Martijn Van Otterlo, and Luc De Raedt. Bellman goes relational. In *International Conference on Machine Learning*, pages 465–472. ACM, 2004.

[85] David Kinny, Magnus Ljungberg, Anand S. Rao, Liz Sonenberg, Gil Tidhar, and Eric Werner. Planned team activity. In Cristiano Castelfranchi and Eric Werner, editors, *Artificial Social Systems, 4th European Workshop on Modelling Autonomous Agents in a Multi-Agent World, MAAMAW '92, S. Martino al Cimino, Italy, July 29-31, 1992, Selected Papers*, volume 830 of *Lecture Notes in Computer Science*, pages 227–256. Springer, 1992. ISBN 3-540-58266-5. doi:10.1007/3-540-58266-5_13.

[86] Alexander Kleiner, Christian Dornhege, and Andreas Hertle. RMASBENCH – rescue multi-agent benchmarking. http://kaspar.informatik.uni-freiburg. de/~rslb (accessed 2012-05-12).

[87] C. E. Koksal and H. Balakrishnan. Quality-Aware Routing Metrics for Time-Varying Wireless Mesh Networks. *IEEE Journal on Selected Areas in Communications*, 24(11):1984–1994, November 2006. ISSN 0733-8716. doi:10.1109/JSAC.2006.881637.

[88] Robert A. Kowalski and Marek J. Sergot. A logic-based calculus of events. *New Generation Computing*, 4(1):67–95, January 1986. ISSN 0288-3635. doi:10.1007/BF03037383.

[89] M. R. Krom. The decision problem for a class of first-order formulas in which all disjunctions are binary. *Mathematical Logic Quarterly*, 13(1-2): 15–20, 1967. ISSN 1521-3870. doi:10.1002/malq.19670130104.

[90] H. W. Kuhn. The Hungarian method for the assignment problem. *Naval Research Logistic Quarterly*, 2:83–97, 1955.

[91] Thomas H. Labella, Marco Dorigo, and Jean-Louis Deneubourg. Division of labor in a group of robots inspired by ants' foraging behavior. *ACM Transactions on Autonomous and Adaptive Systems (TAAS)*, 1:4–25, September 2006. ISSN 1556-4665. doi:10.1145/1152934.1152936.

[92] Jérôme Lang. Belief update revisited. In *Proceedings of the 20th international joint conference on Artifical intelligence*, pages 2517–2522, San Francisco, CA, USA, 2007. Morgan Kaufmann Publishers Inc.

[93] Nuno Lau, Luis Seabra Lopes, Gustavo Corrente, Nelson Filipe, and Ricardo Sequeira. Robot team coordination using dynamic role and positioning assignment and role based setplays. *Mechatronics*, 21(2):445–454, 2011. ISSN 0957-4158. doi:10.1016/j.mechatronics.2010.05.010. Special Issue on Advances in intelligent robot design for the Robocup Middle Size League.

[94] Kristina Lerman, Chris Jones, Aram Galstyan, and Maja J Matarić. Analysis of dynamic task allocation in multi-robot systems. *The International Journal of Robotics Research*, 25:225–241, March 2006. ISSN 0278-3649. doi:10.1177/0278364906063426.

[95] V. Lesser, K. Decker, T. Wagner, N. Carver, A. Garvey, B. Horling, D. Neiman, R. Podorozhny, M. Nagendra Prasad, A. Raja, R. Vincent, P. Xuan, and X. Q. Zhang. Evolution of the GPGP/TÆMS Domain-Independent Coordination Framework. *Autonomous Agents and Multi-Agent Systems*, 9(1-2):87–143, July 2004. ISSN 1387-2532. doi:10.1023/B:AGNT.0000019690.28073.04.

[96] Hector J. Levesque. Planning with Loops. In *Proceedings of the International Joint Conference on Artificial Intelligence (IJCAI)*, pages 509–515, 2005.

[97] Hector J. Levesque, Philip R. Cohen, and José H. T. Nunes. On Acting Together. In *Proceedings of AAAI-90*, pages 94–99, Boston, MA, 1990.

[98] Hector J. Levesque, Raymond Reiter, Yves Lespérance, Fangzhen Lin, and Richard B. Scherl. Golog: A logic programming language for dynamic domains. *Journal of Logic Programming*, 31(1–3):59–83, 1997. doi:10.1016/S0743-1066(96)00121-5.

[99] Martin Lötzsch, Max Risler, and Matthias Jüngel. XABSL - A pragmatic approach to behavior engineering. In *Proceedings of IEEE/RSJ International Conference of Intelligent Robots and Systems (IROS)*, pages 5124–5129, Beijing, China, 2006.

[100] John McCarthy and Patrick J. Hayes. Some Philosophical Problems from the Standpoint of Artificial Intelligence. In B. Meltzer and D. Michie, editors, *Machine Intelligence 4*, pages 463–502. Edinburgh University Press, 1969.

[101] D Mills, J Martin, J Burbank, and W Kasch. Rfc 5905 - network time protocol. *Internet Engineering Task Force IETF*, pages 1–111, 2010. URL http://www.rfc-editor.org/info/rfc5905.

[102] Matthew W. Moskewicz, Conor F. Madigan, Ying Zhao, Lintao Zhang, and Sharad Malik. Chaff: Engineering an efficient sat solver. In *Annual ACM IEEE Design and Automation Conference*, pages 530–535. ACM, 2001.

[103] Sreerama K. Murthy. Automatic construction of decision trees from data: A multi-disciplinary survey. *Data Mining and Knowledge Discovery*, 2: 345–389, 1997.

[104] Ranjit Nair, Milind Tambe, and Stacy Marsella. Role allocation and reallocation in multiagent teams: towards a practical analysis. In *Proceedings of the second international joint conference on Autonomous agents and multiagent systems*, AAMAS '03, pages 552–559, New York, NY, USA, 2003. ACM. ISBN 1-58113-683-8. doi:10.1145/860575.860664.

[105] Alexander Nareyek. *Constraint-Based Agents*, volume 2062 of *Lecture Notes in Computer Science*. Springer Berlin / Heidelberg, 2001. ISBN 978-3-540-42258-7. doi:10.1007/3-540-45746-1.

[106] Dana Nau, Okhtay Ilghami, Ugur Kuter, J. William Murdock, Dan Wu, and Fusun Yaman. Shop2: An htn planning system. *Journal of Artificial Intelligence Research*, 20:379–404, 2003.

[107] S. Nema, John Yannis Goulermas, G. Sparrow, and Phil Cook. A hybrid particle swarm branch-and-bound (hpb) optimizer for mixed discrete nonlinear programming. *IEEE Transactions on Systems, Man, and Cybernetics, Part A*, 38(6):1, 2008. doi:10.1109/TSMCA.2008.2003536.

[108] Salam Nema. *Hybrid evolutionary techniques for constrained optimisation design*. PhD thesis, University of Liverpool, 2010.

[109] J. Neyman. On a New Class of Contagious Distributions, Applicable in Entomology and Bacteriology. *The Annals of Mathematival Statistics*, 10: 35–57, 1939.

[110] Trung Thanh Nguyen and Xin Yao. Benchmarking and solving dynamic constrained problems. In *Proceedings of the IEEE Congress on Evolutionary Computation*, CEC'09, pages 690–697, Piscataway, NJ, USA, 2009. IEEE Press. ISBN 978-1-4244-2958-5. doi:10.1109/CEC.2009.4983012.

[111] Shan-Hwei Nienhuys-Cheng and Ronald de Wolf. *Foundations of Inductive Logic Programming*. Springer-Verlag New York, Inc., Secaucus, NJ, USA, 1997. ISBN 3540629270.

[112] Robert Nieuwenhuis, Albert Oliveras, and Cesare Tinelli. Solving SAT and SAT Modulo Theories: From an abstract Davis–Putnam–Logemann–Loveland procedure to DPLL(T). *J. ACM*, 53:937–977, November 2006. ISSN 0004-5411. doi:10.1145/1217856.1217859.

[113] N. J. Nilsson. *Principles of Artificial Intelligence*. Morgan Kaufmann, San Francisco, CA, USA, 1980.

[114] Thomas Nitsche and Thomas Fuhrmann. A tool for raytracing based radio channel simulation. In *Proceedings of the 4th International ICST Conference on Simulation Tools and Technique*, March 2011.

[115] Hyacinth S. Nwana. Software agents: An overview. *Knowledge Engineering Review*, 11:205–244, 1996.

[116] Object Management Group. OMG Unified Modeling Language™ (OMG UML), Superstructure, May 2010. URL http://www.omg.org/spec/UML/2. 3/Superstructure/PDF/. OMG Document Number: formal/2010-05-05.

[117] Boon Hua Ooi and Aditya K. Ghose. Constraint-based agent specification for a multi-agent stock brokering system. In *Proceedings of the 12th international conference on Industrial and engineering applications of artificial intelligence and expert systems: multiple approaches to intelligent systems*, IEA/AIE '99, pages 409–419, Secaucus, NJ, USA, 1999. Springer-Verlag New York, Inc. ISBN 3-540-66076-3. URL http://dl.acm.org/citation.cfm? id=341506.341611.

[118] Lynne E. Parker. Alliance: An architecture for fault tolerant multi-robot cooperation. *IEEE Transactions on Robotics and Automation*, 14:220–240, 1998.

[119] David Payton, Mike Daily, Regina Estowski, Mike Howard, and Craig Lee. Pheromone robotics. *Autonomous Robots*, 11:319–324, 2001. ISSN 0929-5593. doi:10.1023/A:1012411712038.

[120] Judea Pearl. *Heuristics – intelligent search strategies for computer problem solving*. Addison-Wesley series in artificial intelligence. Addison-Wesley, 1984. ISBN 978-0-201-05594-8.

[121] Adrian Petcu. *A Class of Algorithms for Distributed Constraint Optimization*. Phd. thesis no. 3942, Swiss Federal Institute of Technology (EPFL), Lausanne (Switzerland), October 2007. URL http://liawww.epfl.ch/Publications/Archive/Petcu2007thesis.pdf.

[122] G. D. Plotkin. A Structural Approach to Operational Semantics. Technical Report DAIMI FN-19, University of Aarhus, University of Aarhus, 1981.

[123] Armand E. Prieditis. Machine discovery of effective admissible heuristics. In *Machine Learning*, pages 117–141, 1993.

[124] Patrick Prosser. Hybrid Algorithms for the Constraint Satisfaction Problem. *Computational Intelligence*, 9:268–299, 1993.

[125] D. Pynadath and M. Tambe. Multiagent teamwork: Analyzing the optimality and complexity of key theories and models. In *Proceedings of the 1st conference of autonomous agents and multiagent systems (AAMAS-2002)*, 2002. URL http://citeseer.ist.psu.edu/article/pynadath02multiagent.html.

[126] David V. Pynadath, Milind Tambe, and Nicolas Chauvat. Toward team-oriented programming. In *Intelligent Agents VI: Agent Theories, Architectures, and Languages*, pages 233–247, 1999.

[127] Morgan Quigley, Ken Conley, Brian P. Gerkey, Josh Faust, Tully Foote, Jeremy Leibs, Rob Wheeler, and Andrew Y. Ng. Ros: an open-source robot operating system. In *ICRA Workshop on Open Source Software*, 2009.

[128] J. R. Quinlan. Induction of decision trees. *Machine Learning*, 1:81–106, March 1986. ISSN 0885-6125. doi:10.1023/A:1022643204877.

[129] Anand S. Rao. AgentSpeak(L): BDI agents speak out in a logical computable language. In *MAAMAW '96: Proceedings of the 7th European workshop on Modelling autonomous agents in a multi-agent world : agents breaking away*, pages 42–55, Secaucus, NJ, USA, 1996. Springer-Verlag New York, Inc. ISBN 3-540-60852-4.

[130] Anand S. Rao, Michael P. Georgeff, and E. A. Sonenberg. Social plans: A preliminary report. In E. Werner and Y. Demazeau, editors, *Decentralized AI 3 — Proceedings of the Third European Workshop on Modelling Autonomous Agents in a Multi-Agent World (MAAMAW-91)*, pages 57–76, Kaiserslautern, Germany, 1992. Elsevier Science B.V.: Amsterdam, Netherland. URL http://citeseer.ist.psu.edu/rao92social.html.

[131] Roland Reichle. *Information Exchange and Fusion in Heterogeneous Distributed Environments*. Phd thesis, University of Kassel, Kassel, 2010.

[132] Luís Paulo Reis, Nuno Lau, and Eugenio Oliveira. Situation based strategic positioning for coordinating a team of homogeneous agents. In *Balancing Reactivity and Social Deliberation in Multi-Agent Systems, From RoboCup to Real-World Applications (selected papers from the ECAI 2000 Workshop and additional contributions)*, pages 175–197, London, UK, 2001. Springer-Verlag. ISBN 3-540-42327-3. URL http://dl.acm.org/citation.cfm?id=646142.681100.

[133] Raymond Reiter. The frame problem in the situation calculus: A simple solution (sometimes) and a completeness result for goal regression. In Vladimir Lifschitz, editor, *Artificial Intelligence and Mathematical Theory of Computation: Papers in Honor of John McCarthy*, pages 359–380. Academic Press, San Diego, CA, 1991.

[134] Raymond Reiter. *Knowledge in Action: Logical Foundations for Specifying and Implementing Dynamical Systems*. The MIT Press, Massachusetts, MA, 2001. ISBN 0262182181.

[135] David Richardson. Some Unsolvable Problems Involving Elementary Functions of a Real Variable. *Journal of Symbolic Logic*, 33:5114–520, 1968.

[136] Martin Riedmiller and Heinrich Braun. Rprop - a fast adaptive learning algorithm. *International Symposium on Computer and Information Sciences - ISCIS*, 1992.

[137] RoboCup. RoboCup Foundation. http://www.robocup.org/ (accessed 2012-04-10).

[138] Francesca Rossi, Peter van Beek, and Toby Walsh. *Handbook of Constraint Programming (Foundations of Artificial Intelligence)*. Elsevier Science Inc., New York, NY, USA, 2006. ISBN 0444527265.

[139] Stuart Russell and Peter Norvig. *Artificial Intelligence: A Modern Approach*. Pearson Eduction International, 2nd edition, 2003. ISBN 0-13-080302-2.

[140] Earl D. Sacerdoti. A structure for plans and behavior. Technical Report 109, AI Center, SRI International, 333 Ravenswood Ave., Menlo Park, CA 94025, August 1975.

[141] Sebastian Sardina and Lin Padgham. Goals in the context of BDI plan failure and planning. In *Proceedings of the 6th International Joint Conference on Autonomous Agents and Multiagent systems*, AAMAS '07, pages 7:1–7:8, New York, NY, USA, 2007. ACM. ISBN 978-81-904262-7-5. doi:10.1145/1329125.1329134.

[142] Paul Scerri, David V. Pynadath, Nathan Schurr, Alessandro Farinelli, Sudeep Gandhe, and Milind Tambe. Team oriented programming and proxy agents: The next generation. In Mehdi Dastani, Juergen Dix, and Amal El Fallah-Seghrouchni, editors, *PROMAS*, volume 3067 of *Lecture Notes in Computer Science*, pages 131–148. Springer, 2003. ISBN 3-540-22180-8.

[143] Thomas J. Schaefer. The complexity of satisfiability problems. In *Proceedings of the tenth annual ACM Symposium on Theory of Computing*, STOC '78, pages 216–226, New York, NY, USA, 1978. ACM. doi:10.1145/800133.804350.

[144] Andreas Scharf. Grafische Verhaltensmodellierung kooperativer autonomer Robotersysteme. Bachelor thesis, University of Kassel, Germany, 2008.

[145] Arne Schmitz and Leif Kobbelt. Wave propagation using the photon path map. In *Proceedings of the 3rd ACM international workshop on Performance evaluation of wireless ad hoc, sensor and ubiquitous networks*, PE-WASUN '06, pages 158–161, New York, NY, USA, 2006. ACM. ISBN 1-59593-487-1. doi:10.1145/1163610.1163638.

[146] J. Schubert and H. Sidenbladh. Sequential clustering with particle filters – estimating the number of clusters from data. In *8th International Conference on Information Fusion*, volume 1, pages 1–8. ISIF, July 2005. doi:10.1109/ICIF.2005.1591845.

[147] Bart Selman, Henry A. Kautz, and Bram Cohen. Noise strategies for improving local search. In *Proceedings of the Eleventh National Conference on Artificial Intelligence (AAAI-94)*, pages 337–343, 1994.

[148] M. Senel, K. Chintalapudi, Dhananjay Lal, A. Keshavarzian, and E. J. Coyle. A Kalman Filter Based Link Quality Estimation Scheme for Wireless Sensor Networks. In *Global Telecommunications Conference, 2007. GLOBECOM '07. IEEE*, pages 875–880. IEEE, November 2007. ISBN 978-1-4244-1043-9. doi:10.1109/GLOCOM.2007.169.

[149] Glenn Shafer. *A Mathematical Theory of Evidence*. Princeton University Press, Princeton, 1976.

[150] Murray Shanahan. *Solving the Frame Problem: A Mathematical Investigation of the Common Sense Law of Inertia*. MIT Press, 1997.

[151] Murray Shanahan. The event calculus explained. In *Artificial Intelligence Today: Recent Trends and Developments*, volume 1600 of *Lecture Notes in Computer Science*, pages 409–430. Springer, 1999.

[152] Yi Shang, Markus P.J. Fromherz, and Lara Crawford. A new constraint testcase generator and the importance of hybrid optimizers. *European Journal of Operational Research*, 173(2):419–443, September 2006.

[153] Alex Shtof. Autodiff – high-performance and high-accuracy automatic function-differentiation library suitable for optimization and numeric computing. http://autodiff.codeplex.com/ (accessed 2012-05-21).

[154] J.C. Simon and O. Dubois. Number of solutions to satisfiability instances – applications to knowledge base. *International Journal of Pattern Recognition and Artificial Intelligence*, 3:53–65, 1989.

[155] Hendrik Skubch. Solving non-linear arithmetic constraints in soft realtime environments. In *27th Symposium On Applied Computing*, volume 1, pages 67–75. ACM, 2012.

[156] Hendrik Skubch and Michael Thielscher. Strategy learning for reasoning agents. In João Gama, Rui Camacho, Pavel Brazdil, Alípio Jorge, and Luís Torgo, editors, *Machine Learning: ECML 2005, 16th European Conference on Machine Learning, Porto, Portugal, October 3-7, 2005, Proceedings*, volume 3720 of *Lecture Notes in Computer Science*, pages 733–740. Springer, 2005. ISBN 3-540-29243-8. doi:10.1007/11564096_75.

[157] Hendrik Skubch, Michael Wagner, Roland Reichle, Stefan Triller, and Kurt Geihs. Towards a comprehensive teamwork model for highly dynamic domains. In Joaquim Filipe, Ana L. N. Fred, and Bernadette Sharp, editors, *ICAART 2010 - Proceedings of the International Conference on Agents and Artificial Intelligence, Volume 2 - Agents, Valencia, Spain, January 22-24, 2010*, pages 121–127. INSTICC Press, 2010. ISBN 978-989-674-022-1.

[158] Hendrik Skubch, Daniel Saur, and Kurt Geihs. Resolving conflicts in highly reactive teams. In Norbert Luttenberger and Hagen Peters, editors, *17th GI/ITG Conference on Communication in Distributed Systems, KiVS*, volume 17 of *OASICS*, pages 170–175. Schloss Dagstuhl - Leibniz-Zentrum für Informatik, Germany, 2011. ISBN 978-3-939897-27-9. doi:10.4230/OASIcs.KiVS.2011.170.

[159] Hendrik Skubch, Michael Wagner, Roland Reichle, and Kurt Geihs. A modelling language for cooperative plans in highly dynamic domains. *Mechatronics*, 21(2):423–433, 2011. Special Issue on Advances in intelligent robot design for the Robocup Middle Size League.

[160] Jon Sneyers, Tom Schrijvers, and Bart Demoen. The computational power and complexity of constraint handling rules. *ACM Trans. Program. Lang. Syst.*, 31(2):8:1–8:42, February 2009. doi:10.1145/1462166.1462169.

[161] David Steinberg, Frank Budinsky, Marcelo Paternostro, and Ed Merks. *EMF: Eclipse Modeling Framework 2.0*. Addison-Wesley Professional, 2009. ISBN 0321331885.

[162] Jayoung Sung, Henrik I. Christensen, and Rebecca E. Grinter. Sketching the Future: Assessing User Needs for Domestic Robots. In *The 18th IEEE International Symposium on Robot and Human Interactive Communication*, pages 153–158, 2009. doi:10.1109/ROMAN.2009.5326289.

[163] Milind Tambe. Towards flexible teamwork. *Journal of Artificial Intelligence Research*, 7:83–124, 1997. URL http://citeseer.ist.psu.edu/tambe97towards.html.

[164] Milind Tambe, David V. Pynadath, and Nicolas Chauvat. Building dynamic agent organizations in cyberspace. *IEEE Internet Computing*, 4:65–73, March 2000. ISSN 1089-7801. doi:10.1109/4236.832948.

[165] Russell Taylor and Leo Joskowicz. Computer-integrated surgery and medical robotics. *The Hand*, 2002:1199–1222, 2008.

[166] Michael Thielscher. Introduction to the fluent calculus. *Electronic Transactions on Artificial Intelligence*, 2:179–192, 1998. URL http://www.ep.liu.se/ej/etai/1998/006/.

[167] Michael Thielscher. Flux: A logic programming method for reasoning agents. *Theory and Practice of Logic Programming*, 5:533–565, July 2005. ISSN 1471-0684. doi:10.1017/S1471068405002358.

[168] Michael Thielscher. *Reasoning robots: the art and science of programming robotic agents*. Applied logic series. Springer, 2005.

[169] Marc Toussaint, Nils Plath, Tobias Lang, and Nikolay Jetchev. Integrated motor control, planning, grasping and high-level reasoning in a blocks world using probabilistic inference. In *IEEE International Conference on Robotics and Automation (ICRA)*, 2010.

[170] Douglas Vail and Manuela Veloso. Dynamic multi-robot coordination. In *Multi-Robot Systems: From Swarms to Intelligent Automata, Volume II*, pages 87–100. Kluwer Academic Publishers, 2003.

[171] Pascal Van Hentenryck. Numerica: a modeling language for global optimization. In *Proceedings of the Fifteenth international joint conference on Artifical intelligence - Volume 2*, pages 1642–1647, San Francisco, CA, USA, 1997. Morgan Kaufmann Publishers Inc. ISBN 1-555860-480-4. URL http://dl.acm.org/citation.cfm?id=1622270.1622392.

[172] Pascal Van Hentenryck and Thierry Le Provost. Incremental search in constraint logic programming. *New Generation Computing*, 9:257–275, 1991. ISSN 0288-3635. doi:10.1007/BF03037165.

[173] Gérard Verfaillie and Thomas Schiex. Solution reuse in dynamic constraint satisfaction problems. In *Proceedings of the twelfth national conference on Artificial intelligence (vol. 1)*, AAAI '94, pages 307–312, Menlo Park, CA, USA, 1994. American Association for Artificial Intelligence. ISBN 0-262-61102-3. URL http://dl.acm.org/citation.cfm?id=199288.178066.

[174] Felix von Leitner. The Dark Side of C++. http://www.fefe.de/c++/c%2B%2B-talk.pdf (accessed 2012-05-22), 2007. Presented at Chaos Communication Camp, Berlin, 2007.

[175] Tingting Wang, Jiming Liu, and Xiaolong Jin. Minority game strategies in dynamic multi-agent role assignment. In *Proc. of the IEEE/WIC/ACM International Conference on Intelligent Agent Technology*, pages 316–322, Washington, DC, USA, 2004. IEEE Computer Society. ISBN 0-7695-2101-0. doi:10.1109/IAT.2004.79.

[176] J. H. Ward. Hierarchical Grouping to Optimize an Objective Function. *Journal of the American Statistical Association*, 58(301):236–244, March 1963. ISSN 01621459. doi:10.2307/2282967.

[177] Jörg Weber and Franz Wotawa. Combining runtime diagnosis and ai-planning in a mobile autonomous robot to achieve a graceful degradation after software failures. In Joaquim Filipe, Ana L. N. Fred, and Bernadette Sharp, editors, *ICAART 2010 - Proceedings of the International Conference on Agents and Artificial Intelligence, Volume 1 - Artificial Intelligence, Valencia, Spain, January 22-24, 2010*, pages 127–134. INSTICC Press, 2010. ISBN 978-989-674-021-4.

[178] T. Weigel, J.-S. Gutmann, M. Dietl, A. Kleiner, and B. Nebel. CS Freiburg: Coordinating robots for successful soccer playing. *IEEE Transactions on Robotics and Automation*, 18(5):685–699, 2002.

[179] Terry Winograd. *Understanding Natural Language*. Academic Press, New York, 1972. doi:10.1002/bs.3830180608.

[180] D. H. Wolpert and W. G. Macready. No free lunch theorems for optimization. *Evolutionary Computation, IEEE Transactions on*, 1(1):67–82, April 1997. ISSN 1089-778X. doi:10.1109/4235.585893.

[181] Michael Wooldridge. *An Introduction to MultiAgent Systems*. Wiley Publishing, 2nd edition, 2009. ISBN 0470519460, 9780470519462.

[182] Michael Wooldridge, Nicholas Jennings, and David Kinny. A methodology for agent-oriented analysis and design. *Journal of Autonomous Agents and Multi-Agent Systems*, 3:285–312, 1999.

[183] John Yen, Jianwen Yin, Thomas R. Ioerger, Michael S. Miller, Dianxiang Xu, and Richard A. Volz. Cast: Collaborative agents for simulating teamwork. In *Proceedings of the 17th International Joint Conference on Artificial intelligence - Volume 2*, IJCAI'01, pages 1135–1142, San Francisco, CA, USA, 2001. Morgan Kaufmann Publishers Inc. ISBN 1-55860-812-5, 978-1-558-60812-2.

[184] Jianwen Yin, Michael S. Miller, Thomas R. Ioerger, John Yen, and Richard A. Volz. A knowledge-based approach for designing intelligent team training systems. In Carles Sierra, Maria Gini, and Jeffrey S. Rosenschein, editors, *Proceedings of the Fourth International Conference on Autonomous Agents*, pages 427–434, Barcelona, Catalonia, Spain, 2000. ACM Press. doi:10.1145/336595.337560.

[185] Weixiong Zhang, Guandong Wang, Zhao Xing, and Lars Wittenburg. Distributed stochastic search and distributed breakout: properties, comparison

and applications to constraint optimization problems in sensor networks. *Artificial Intelligence*, 161(1-2):55–87, January 2005. ISSN 0004-3702. doi:10.1016/j.artint.2004.10.004.

[186] H.-J. Zimmermann. *Fuzzy set theory—and its applications (3rd ed.)*. Kluwer Academic Publishers, Norwell, MA, USA, 1996. ISBN 0-7923-9624-3.

[187] Oliver Zweigle, Reinhard Lafrenz, Thorsten Buchheim, Uwe-Philipp Käppeler, Hamid Rajaie, Frank Schreiber, and Paul Levi. Cooperative agent behavior based on special interaction nets. In Tamio Arai, Rolf Pfeifer, Tucker R. Balch, and Hiroshi Yokoi, editors, *Proceedings of the 9th International Conference on Intelligent Autonomous Systems - IAS, University of Tokyo, Tokyo, Japan, March 7-9, 2006*, pages 651–659. IOS Press, 2006. ISBN 1-58603-595-9.